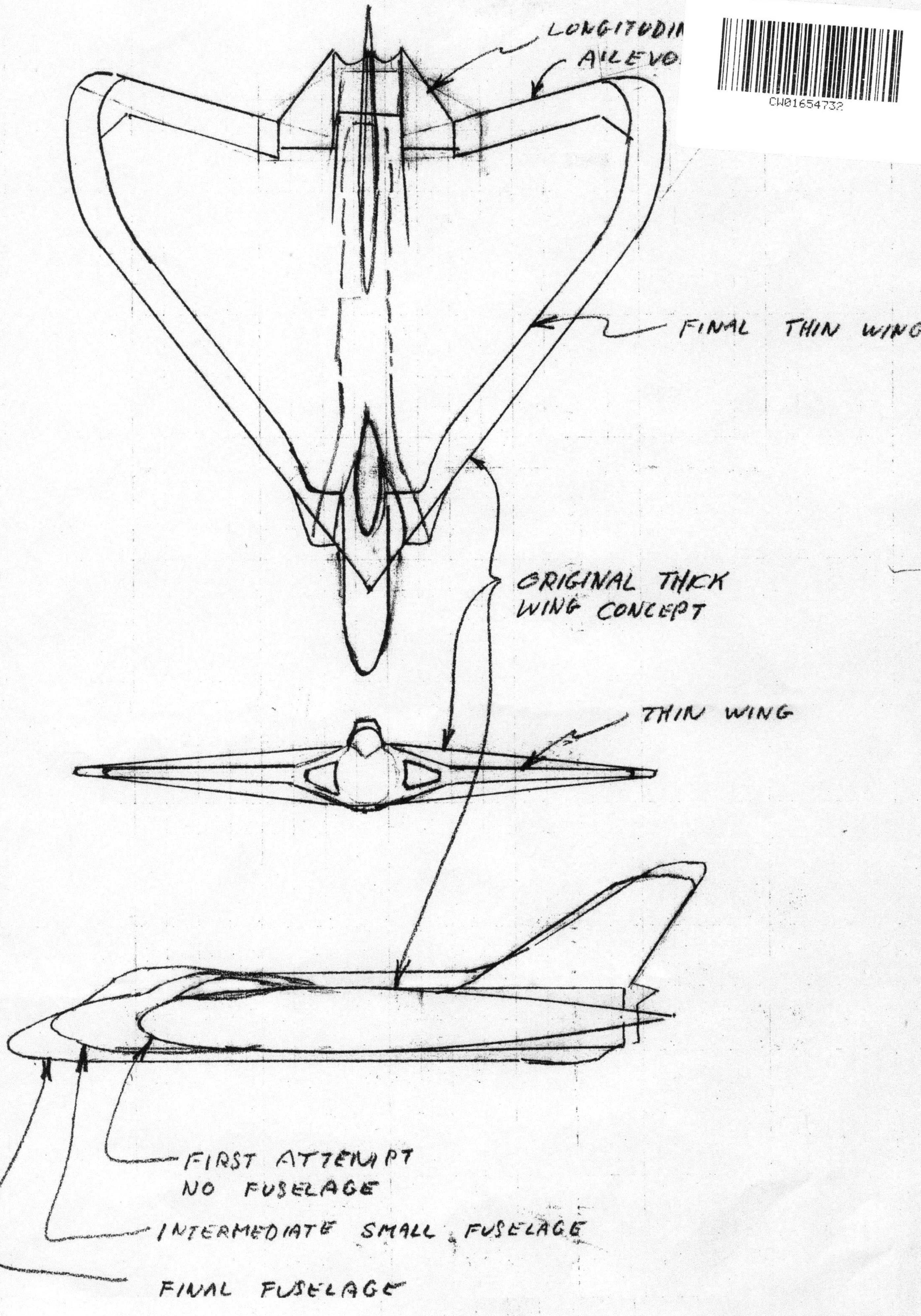

LONGITUDIN
AILEVO
FINAL THIN WING
ORIGINAL THCK
WING CONCEPT
THIN WING
FIRST ATTEMPT
NO FUSELAGE
INTERMEDIATE SMALL FUSELAGE
FINAL FUSELAGE

Killer Rays

Story of the Douglas F4D Skyray and F5D Skylancer

Mark A. Frankel

specialty press
PUBLISHERS AND WHOLESALERS

Specialty Press
39966 Grand Avenue
North Branch, MN 55056
Phone: 651-277-1400 or 800-895-4585
Fax: 651-277-1203
www.specialtypress.com

Edit by Mike Machat
Layout by Monica Seiberlich

ISBN 978-1-58007-155-0
Item No. SP155

Library of Congress Cataloging-in-Publication Data

Frankel, Mark.
Killer rays : the story of the Douglas F4D Skyray and F5D Skylancer / by Mark Frankel.
p. cm.
Includes bibliographical references and index.
ISBN 978-1-58007-155-0
1. Skyray fighter plane--History. I. Title.
UG1242.F5F695 2010
359.9'4834--dc22

2010013768

Printed in China
10 9 8 7 6 5 4 3 2 1

Front Cover:
"Hook down, gear down, in the groove." Moments away from "trapping" aboard the USS Hornet *(CV-8), this F4D-1 from VF-23 nicely displays its classic bat-wing planform while crossing the fantail.* (Arland Nankivil via Mark Nankivil Collection)

Front Flap:
XF4D sits on the ramp at Edwards AFB during initial flight tests conducted at that installation before later extensive Naval testing at NAS Patuxent River, Maryland. The Skyray prototype was painted in an all-white test scheme. (U.S. Navy via Nick Williams Collection)

End Papers:
Original drawings courtesy of the Nick Williams Collection.

Title Page:
Artist Craig Kodera captures the raw power of an intercept scramble as two F4D-1s from VMF(AW)-513 launch into the early morning skies from NAS Atsugi, Japan, in 1961. A majestic Mount Fuji looms in the background (Craig Kodera)

Facing Page:
After being retired from front-line units in 1962, Skyrays found a new lease on life in utility squadrons. Here, F4D-1 BuNo 139122 poses on the ramp at Long Beach in May 1964 after being sent back to Douglas as a demonstrator. (Boeing via Nick Williams Collection)

Back Cover Photos

Top left:
Two F4D-1 Skyrays, their wings still folded for optimum shipboard storage, are seen being moved into position for launch from the deck of the USS Franklin D. Roosevelt *(CV-42) in 1957.* (U.S. Navy photo)

Top right:
Famed Douglas test pilot Bob Rahn (right) congratulates Navy LCDR Jim Verdin after he set the world absolute speed record in the XF4D-1 seen in the background. Rahn helped prepare the Skyray for its record run. (Nick Williams Collection)

Bottom:
Although never put into production, the advanced Douglas F5D Skylancer served as a valuable research aircraft for NASA. Here, NASA 213 prepares for takeoff at Edwards AFB, home of the Dryden Flight Research Facility. (NASA DFRC photo)

Distributed in the UK and Europe by
Crécy Publishing Ltd
1a Ringway Trading Estate
Shadowmoss Road
Manchester M22 5LH England
Tel: 44 161 499 0024
Fax: 44 161 499 0298
www.crecy.co.uk
enquiries@crecy.co.uk

TABLE OF CONTENTS

Foreword

by Nick Williams

During the early 1950s, I was convinced that I was one of the luckiest boys in the world. I was born and raised in El Segundo, California, living just a quarter-mile from the main runways at Los Angeles International Airport. Prop-driven airliners departed nearly every minute, along with the new jet fighters and attack aircraft coming off the nearby production lines at North American Aviation and Douglas Aircraft Company, where my father had worked since 1937 when it was known as the Northrop Division of Douglas. Before I was old enough to ride my bicycle down Imperial Highway, which bordered the airport to the south, I often could be found sitting on top of our garage with a pair of binoculars watching the departing aircraft. The special days were when a strong, cold wind blew in from the north and the landing pattern was shifted to Los Angeles International's short north-south runway. The airliners would approach over residential El Segundo, descending above our back yard close enough for me to make out the passengers' faces peering out as I waved to them like a fool while perched on my garage.

Once I was trusted enough to ride my bike down Imperial Highway to Sepulveda Boulevard, I spent many Saturdays peddling to the commercial airline terminal buildings on the north side of the airport. Climbing up to the observation deck, I'd mostly watch the United, American, TWA and Western Airlines Convairliners, DC-3s, DC-6s, and Constellations load and unload passengers. If I was feeling especially adventurous, I would ride eastward on Imperial to Aviation Boulevard to stand along the railroad tracks that ran in front of the twin runways of Los Angeles International, where you felt you could almost touch the wheels of landing aircraft as they roared overhead. Of course, there were signs posted along the airport's perimeter fence prohibiting you from loitering as well as taking pictures. More than a few times the local police or airport security patrols would politely ask me to leave, but that was a small price to pay for the experience of being the luckiest boy in the world.

While I was able to see firsthand Douglas El Segundo's products at their annual open house—various models of the AD Skyraider, the F3D Skyknight and the new A4D Skyhawk and A3D Skywarrior—my first vivid recollection of the F4D-1 Skyray was perhaps in 1954 while standing on these railroad tracks on Aviation Blvd. I had been accustomed to watching North American F-100s rumble up to Runway 25L and take off with a loud *crack* of their afterburners...exhilarating to a 10-year-old airplane nut. But when a dark blue Skyray came scuttling down the taxiway, it was as if a flying saucer was approaching. Whereas the F-100, like most fighter aircraft of the period, was all angles and sharp profiles, the Skyray was beautifully curved with flowing lines from the radome to its swept tail. It looked like nothing else. I remember standing in awe as the Skyray's engine revved up with a belch of smoke, its pilot released the brakes, and the afterburner kicked in with a loud *bam*! that literally could be felt. In what seemed like only a few seconds, the Skyray lifted off the shimmering runway, tucked in its landing gear, and streaked nearly straight up on a tail of flame as it disappeared into the coastal haze. In my mind's eye, I can see myself standing there breathing in the burnt JP4, simply awestruck at what I had just witnessed.

Over the following 30 years, I scooped up photographs and every tidbit of information on the Skyray I could beg or borrow from the Douglas PR folks, local newspapers, and eventually from many of the men directly involved in the F4D-1's design, production, and service life. The late Douglas historian, author, and photographer Harry Gann was instrumental in opening doors for me at Douglas as well as leading me to the whereabouts of many of the Skyray's design team and former pilots. Between 1970 and 1984, I was fortunate to have lengthy face-to-face interviews with engineers Ed Heinemann, A.M.O. Smith, and R. G. Smith; Douglas Testing Division pilots George Jansen, Bob Rahn, Jim Stegman, and Walt Harper; NATC Director of Flight Test Marshall Beebe; former VF-51 Skyray pilot Bill Davis and General Electric test pilot Bill Todd. Most of these men graciously invited me into their homes to reminisce with them about the Skyray. Jim Stegman and Walt Harper were interviewed at Douglas' Palmdale, California, facility and I spoke with R. G. Smith in his office at Douglas Long Beach. Those pilots whom I couldn't interview directly (on a teacher's budget) patiently answered my questions with detailed letters: Robert E. Drew (Douglas Testing Division), time-to-climb record flights pilot Col. Ed LeFaivre, C. A. Jorgenson (VMF(AW)-115), Roy S. Johnston (VF-51), H. E. Swanson (VF-13), Dudley Gillaspy (VF-882), and NATC Skyray test and evaluation pilots Tim Wooldridge, George Watkins, Gerald O'Rourke, and Phil Oestreicher.

While researching the Skyray's history and writing about its service life, several questions concerning its proper place in American aviation history arose that have remained unanswered, due either to my laziness or the demands of other aircraft research projects. How proficient was the Skyray in fulfilling its mission? How did the Skyray affect the fortunes of Naval Aviation? How did it compare to other contemporary fighter aircraft in flying qualities as well as accident rates? I am happy to say that these, as well as other questions that have gone unanswered for more than 50 years, are now addressed in Mark Frankel's fine book on the Skyray.

Former Skyray pilots have commented on its merits—and faults—in wide-ranging opinions ranging from "The worst Dutch-roller ever to fly" and "Don't ever buy another one of these" airplanes, to "The best fighter aircraft at altitude" and "A very capable airplane if one is willing to learn to fly it properly." Historians can always come up with "What ifs" when judging an airplane's place in aviation history. What if the Skyray had been equipped with the J57 from the outset? What if the F4D-2/F5D-1 had been placed in quantity production instead? For myself, I feel that these ruminations accomplish nothing. However, the following unarguable facts can't be ignored when assessing the Skyray's actual accomplishments, while acknowledging its idiosyncrasies: 1) Two Air Force Evaluation teams flew the airplane at Edwards Air Force Base (AFB) in December 1953 and July 1955, reportedly in a vain attempt to find an Air Force fighter that could outmaneuver the Skyray at altitude; 2) The XF4D1's 3-km world's airspeed record—not insignificantly the first to be set by a carrier-based airplane—could not be broken by the J57-powered YF-100 by the required 1 percent, so the Air Force was forced to use a technicality to set a new record at 15 km. The Skyray's 3-km record, set in 1953, wasn't broken until 1961 by a Navy McDonnell F4H Phantom II ; 3) The Skyray's 100-km closed course record, flown below 100 meters, lasted for over five years until bested in 1959 by a Dassault Mirage III, flying at an altitude of 22,970 feet; 4) Edward H. Heinemann shared the 1953 Collier Trophy with North American Aviation's James H. Kindelberger. Awarded each year for the greatest aviation achievement in America, the two men accepted the trophy for designing the first supersonic airplanes to enter service: the F-100 Super Sabre and the Douglas F4D-1 Skyray; 5) While the Skyray's time-to-climb records, flown mid-way in its service life, lasted less than a year before being bested by the Air Force's F-104, it should be pointed out that not only did the F4D-1 set an entirely new climb record to 15,000 meters (nearly 50,000 feet) but that at the time (1958) the Douglas Skyray still held seven world's records. Not a bad legacy for an airplane designed in the late 1940s to be the U.S. Navy's first carrier-based interceptor.

Nicholas M. Williams
Aviation Historian
Waverly, Iowa

ACKNOWLEDGMENTS

This book could not have been written without the brilliant research of Nicholas M. Williams (better known as Nick). Born into a Douglas family during the peak of World War II production, Nick grew up among the engineers, test pilots, production line workers, and factory executives that were responsible for the most exciting Naval aircraft program of the 1950s—the F4D-1 Skyray. From childhood, Nick started collecting photographs, technical drawings, and reports on the development, production, and service history of the Skyray. He cultivated relationships with many of the key players in the Skyray story including designer Ed Heinemann, aerodynamicist A.M.O. Smith, and test pilot Bob Rahn. They granted him interviews in their homes (which he wisely taped), provided him with documents from their personal files, and corresponded with him as questions arose from his continuing research. Nick authored a series of articles on the F4D that appeared in the respected *Journal of the American Aviation Historical Society* from 1976 to 1986. This material was expanded and republished in Steve Ginter's Naval Fighter series of monographs. Nick had become known as the leading authority on the Douglas Skyray and I approached him in that capacity when I was doing some work on the F4D-1 in the early 1990s. I found Nick not only extremely knowledgeable but also extremely generous with his collection of photographs and documents, many of which were priceless. When I considered writing this book Nick gave me access to his entire collection of material that had been accumulating for more than 50 years. I doubt that any author has been given the gift of such exquisitely detailed research in one neat bundle. Nick has done a commendable job of preserving the "Tribal Knowledge" of the Douglas Skyray, and I am privileged to stand on his shoulders.

When it comes to perfect knowledge of aircraft colors and markings I rely on Jack Morris. While writing the chapters on F4D-1 Fleet service it occurred to me that Jack's renderings would be a perfect way to illustrate the diversity of schemes worn by the *Ford* during this colorful period in Naval Aviation. I commissioned Jack to produce a dozen or so illustrations showing the markings of the major Skyray operators, but Jack wildly exceeded my expectations by covering nearly every unit, and in many cases illustrating the evolution in unit markings for specific cruises or mission changes. The most impressive thing about Jack's work is his attention to detail and concern for color accuracy. In many cases the source data was poor, usually just a black and white photograph, but Jack was able to construct a definitive color record of the Skyray and Skylancer.

Mark Nankivil was another valuable source of Skyray lore. Mark's father, Arland, flew Fords for VF-23 aboard the Hancock and was selected for VF(AW)-3, the elite NORAD squadron. Mark has collected Skyray memorabilia for most of his life and has become a respected aviation historian. His interests range from contemporary military aircraft to vintage sailplanes and his photo collection is notable. Mark introduced me to his father who told some fascinating "sea stories" about Ford scrambles from North Island. Mark also led me to another accomplished aviation photographer, Stephen Miller, who supplied numerous Skyray and Skylancer images.

I am deeply indebted to Bill Larkins, Rob Getz, Don Spering, Robert Dorr, and Ron Piccianni who opened their extensive photo collections to me, and I am particularly grateful to Alan Renga of the San Diego Air & Space Museum for his rapid responses to my frequent cries for help with F-102 and F-106 material.

I've learned that historic research is like criminal investigation—one clue leads to another. Kimble McCutcheon, president of the Aircraft Engine Historical Society, was a valuable resource for my chapter on the Westinghouse J40. He not only provided images of various jet engines from the Society's archives, but he introduced me to Paul Lagasse's scholarly paper on the Westinghouse Aviation Gas Turbine Division, which explained the entire J40 episode in great detail. Paul, in turn, informed me of the Westinghouse photograph collection at the Hagley Museum and Library. Jon Williams and Judy Stevenson at Hagley allowed me to study hundreds of photographs and scan the most relevant for the Skyray story.

CDR Doug Siegfried of the Tailhook Association is always a good source for photographs of Naval Aviation. He confirmed this fact with a disc of Skyray photos, many never published before. Some of these valuable shots depict rare squadron markings that had been hard to document. Former Skyray pilots, Gary McClellan, Arland Nankivil, Joel Premselaar, Phil Oestreicher, Rod Creagh, and James Campbell provided firsthand accounts of their experiences along with personal photographs. Their stories added a welcomed dose of human interest and color to an otherwise dry recitation of the historic record.

Bud Southworth found and photographed the F5D-1 Skylancer that was used for the Ogee wing tests at NASA Ames. I knew of the Skylancer at the Neil Armstrong Museum, and I assumed that it was the sole survivor. Bud tracked down this second airframe, which has been preserved in excellent condition, and spent two days photographing it in great detail.

Finally, I am deeply indebted to noted aviation historian and author, Tommy Thomason. He has been enormously helpful on a wide range of issues from explanations of obscure engineering points (the rotating compressor stall that afflicted the J40 is an example) to explanations on the fundamental principals of Naval Aviation (the aircraft procurement process of BuAer in the 1950s is an example). Tommy provided material that was second only to Nick Williams in volume, but equally important in content. The entire Board of Inspection and Survey (BIS) report for the Skyray Acceptance Trials, as well as the mock-up reports for the F4D-1 and F4D-2 (later the F5D-1) came from the Thomason collection.

I was born at the end of World War II and I became aware of—no, obsessed with—aviation at the time the Skyray was being developed. Like Nick Williams, every time I caught sight of a Ford at one of the airfields I frequented (NAS Willow Grove, NADC Johnsville, or NATF Lakehurst to name a few) it stopped me in my tracks. Over the years I have read everything on the Skyray that I could get my hands on. Recently it occurred to me that the Skyray was more important to Naval Aviation than many people realize. With the help of authorities like Nick Williams and Tommy Thomason I set out to examine the entire record. I am still frustrated by confusion over certain facts. I am uncertain of which XF4D-1 was first to fly above Mach 1 in 1952—I think it was 586 once the non-afterburning J40 was installed. I am also uncertain of whether 586's wing had operating slats during its first flight in 1951. Nevertheless, it has been enormously rewarding to research and recite this body of historic data on an airplane considered to be the Navy's first supersonic fighter.

Mark A. Frankel
Gladwyne, Pennsylvania
August 2010

DEDICATION

This book is dedicated to the loving memory of author Mark Frankel's wife, Jill (1946–2010), and illustrator Jack Morris's wife, Dawn (1941–2009), both of whom passed away during the creation of this book.

ABOUT THE AUTHOR

Mark Frankel's life-long interest in aviation began as a flying-aircraft modeler in early childhood. As a teenager, he competed in Academy of Model Aeronautics meets at the Nationals level, which led to his interest in Naval Aviation. After completing law school, Mark served as a trial lawyer in Navy JAGC and earned his FAA Private License as well.

Today, Frankel remains an active pilot flying both radio-controlled models and his own Piper PA-28 Archer. He has written numerous articles on flying models, and authored a book on the Navy Temco TT-1 Pinto. *Killer Rays* is Mark's first book for Specialty Press, and is the ultimate outcome of his initial fascination with the Douglas F4D Skyray dating back to his days flying radio-controlled models at Naval Air Station Willow Grove, Pennsylvania.

INTRODUCTION

"It is appearances, characteristics, and performance that make a man love an airplane, and they, told truly, are what put emotion into one. You love a lot of things if you live around them, but there isn't any woman and there isn't any horse, not before nor any after, that is as lovely as a great airplane, and men who love them are faithful to them even though they leave them for others."

— Ernest Hemingway

The Skyray generated a lot of emotion. It was an elegant solution to a defense requirement that was, at the time, impossible to accomplish. It was meant to climb like a bat out of hell and fly faster than anything else in the air—it accomplished both. However, it took a long time to develop an airplane with these seemingly impossible abilities and newer models eclipsed it with more elegant solutions by the time it entered front line service. The Skyray was unusual; there was nothing else like it in Naval Aviation and it took a lot to learn its quirks. Some pilots loved it, and considered flying it the high point of their careers. Others loathed it and even refused to fly it.

My purpose in writing this book was to provide the reader with an explanation of how and why the Skyray came to be, to examine the technological problems that it overcame, to analyze its service history, to recreate how it was flown, to study its stillborn successor the Skylancer, and to explore its place in Naval Aviation history with the advantage of 50 years of hindsight.

Along the way I wanted to examine some issues that remain, to this day, matters of debate. Was the Skyray truly supersonic? The answer is an unequivocal maybe. How did the Air Force recapture the world's speed record? The answer is with slight-of-hand. Why did the Navy stick with the Westinghouse J40 engine for so long? I still don't know the answer to that one.

While writing about the evolution of the Navy's first delta-winged fighter it occurred to me that the Air Force was going through the same exercise in a parallel universe with the Convair F-102. It would be interesting, I thought, to examine that program and point out the similarities and differences. In each chapter, where it is relevant, a sidebar describes what was going on with the Air Force's program. When it came to analyzing the Skylancer, the F-106 was emerging on the Air Force side.

One difference that became readily apparent between the Air Force and Navy interceptor programs is that the Skyray development was relatively linear. That is to say, the path from the prototype XF4D-1 to the final production F4D-1 was straight. The Skyray didn't change that much. The Air Force program beginning with the XP-92 was fraught with roadblocks and detours. It was truly a long and winding road from the XP-92 to the F-102A.

I hope you enjoy reading about this unique period in aviation history when technology was advancing at blinding speed, defense planners were trying to respond to hypothetical threats, and defense contractors were promising performance that they often failed to deliver.

I am reminded of a famous French adage, "The more things change, the more they remain the same."

CHAPTER ONE

ORIGINS OF THE SKYRAY

Retouched photography was an industry staple during the Cold War era, and Douglas's Art Department proudly rallied to the cause in this heavily—and noticeably—retouched photo of the XF4D-1 prototype. Notice the razor-sharp edges and gray shading on critical areas of the airframe that had to be kept classified. (Boeing via Nick Williams Collection)

The F4D-1 Skyray (or "Ford" as it was affectionately nicknamed) was noted for its astonishing rate of climb. For a time it was the fastest airplane in the world, and its designer, Edward H. Heinemann, was awarded the Collier Trophy by President Eisenhower for the greatest achievement in American aviation. Yet the Douglas Skyray never fired a shot in anger. It took more than five years to become operational, and it served with the Fleet for just eight years. Only 421 Skyrays were built, and on occasion it was tasked with missions beyond its design requirements.

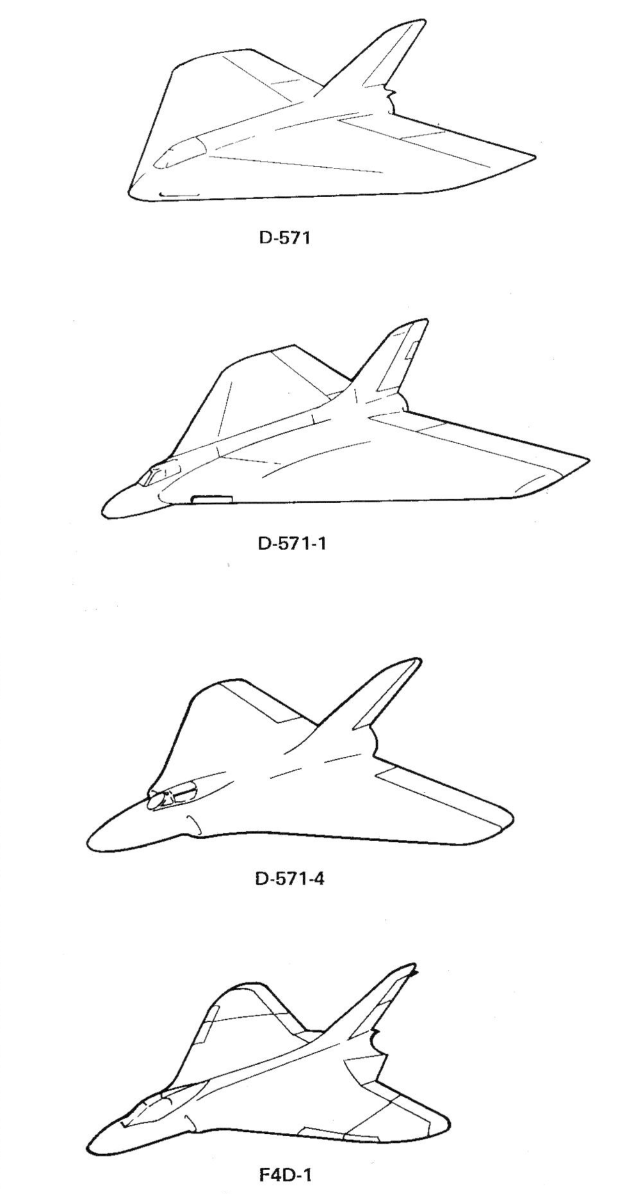

Illustrations of the Model D-571 tailless concept from Douglas interoffice memos between Heinemann and his design team. The illustrations show the evolution of shape from October 1946 to the initial BuAer contract of June 1947. (Boeing via Nick Williams Collection)

But the Skyray provided the Navy with its first truly high-performance jet aircraft and yielded a wealth of experience for the next generation of proud Navy fighters—the F-8 Crusader and F-4 Phantom—which employed lessons learned from the Skyray with great distinction. The origins of the Skyray can be traced to the late 1930s when several influential American aeronautical engineers began to show interest in tailless designs as a means of reducing drag and improving efficiency. The recognized leader in the tailless movement was John K. Northrop, the founder of Northrop Corporation of El Segundo, California. At that time Northrop's company was a joint venture with the Douglas Aircraft Company.

In 1932 Northrop hired Ed Heinemann, a talented self-trained engineer, who worked with Northrop on a tailless design designated the Model 25. The design produced promising wind tunnel data, but the lack of Douglas corporate interest precluded building a full-sized example. Jack Northrop was so committed to the exploration of tailless concepts, that he resigned from the Douglas/Northrop joint venture and formed an independent company. Douglas assumed control of the former Northrop company, renamed it the El Segundo Division of Douglas Aircraft, and appointed Ed Heinemann chief engineer.[1] Douglas, El Segundo flourished under Heinemann, producing a series of conventional attack aircraft such as the SBD Dauntless, TBD Devastator, and BT2D-1/AD Skyraiders. The tailless concept remained dormant at Douglas during the war years.

In 1939, shortly after Northrop's departure, Heinemann hired a recent graduate from Caltech, Apollo Milton Olin Smith. Known to his friends as A.M.O. (or Amo), Smith proved to be a brilliant aerodynamicist with an uncanny ability to predict aircraft performance accurately from wind tunnel data. In 1945, at the end of the war in Europe, the War Department invited several noted engineers and scientists to examine captured German documents. Heinemann's aerodynamicists, Eugene Root, and A.M.O. Smith were selected to join the technical mission. Smith recalled, "Root and I did not know what we could do or be assigned to do, but very quickly we found a useful job. At Göttingen, one of the major German aeronautical research facilities, when they found they were losing the war, they threw all their classified reports down a well. Our Army discovered them, recovered them, and brought them to Paris to dry...Our job was to index this material, copy it, and get it to the U.S. fast because the war with Japan was still going on."[2]

The reports that Root and Smith studied contained provocative discoveries about high-speed flight. It was apparent that the Germans were experimenting with concepts that could advance aircraft performance dramatically. Smith recalled, "Easily the most startling we found was the swept wing. Except for some preliminary theory by R. T. Jones, no one in the U.S. had thought of swept wings, and of course no tests had been done. The Germans had tested hundreds of models."

Smith also became enthused about the possibilities of tailless aircraft after attending several lectures by Dr. Alexander Lippisch, the designer of the Me 163 Komet, the only tailless aircraft to see operational service in World War II. These lectures exposed Smith to early theory on delta-wing configurations, and when he returned to Douglas, El Segundo, he was eager to put these theories into practice. "Based a good deal on what I had learned in Europe, it looked to me like a very good design could be made out of a low-aspect ratio, swept back, tailless airplane. It had a number of advantages that the higher aspect ratio tailless airplanes like Northrop's lacked. Accordingly, one day I submitted a memo to Heinemann with a sketch of a proposed wing planform. He got interested, and I handled preliminary wind tunnel tests of the wing I suggested...the tests turned out well."[3]

Heinemann, by now, had assembled a highly talented design team of Leo J. Devlin, chief designer; R. G. Smith, configuration engineer; and A.M.O. Smith, aerodynamicist. They spent much of 1945 and 1946 trying to apply the Lippisch theories and wind tunnel data to a new fighter design. Douglas assigned the designation Model D-571 to the project. The earliest iteration of Model D-571 consisted of a 700 square foot flying wing with no separate fuselage; the cockpit and engine bay were simply faired into the wing center section. The design team was concerned by the thickness of the center section, which was required to house the engine, cockpit, fuel cells, and other internal components. Wind tunnel data confirmed that it would seriously degrade transonic performance. Gradually it was realized that a conventional fuselage was necessary and that the wing had to be thin.

A breakthrough occurred when it was decided to blend the fuselage into a thin wing using a large tapered fillet. This novel blended shape is still employed on many contemporary military aircraft; the F-16, B-1, and SR-71 are a few examples. As recently as 2009 blended wing/body configurations were being explored by NASA for high efficiency airliners. Gradually, R. G. Smith refined the shape into a variant designated D-571-4. It had a wing area of 600 square feet and an estimated gross weight of 16,000 pounds.

International events were about to affect Heinemann's design team. In August 1947 three Tupolev Tu-4 Bull heavy bombers were revealed by the Soviet Union at the Tushino Air Display. Suddenly western defense planners realized the possibility of a Soviet long-range assault. Air Force Secretary W. Stuart Symington III appointed a board of senior Air Force officers to select a weapon system for this scenario. The Navy also began to form a strategy in response to the Soviet long-range threat, and focused attention on the Douglas D-571 tailless project. However, the Navy's Bureau of Aeronautics (BuAer), the organization responsible for the design, development, procurement, and support of Naval aircraft, was highly conflicted. During World War II BuAer had been successful providing the Fleet with an endless stream of state of the art airplanes, but these were conventional propeller driven aircraft with time-tested aerodynamic features. Now BuAer was tasked with entering the jet age, and responding to threats that were often theoretical. At first BuAer's approach was profoundly cautious. Jets presented enormous problems in naval aircraft

A rendering of the proposed Model D-571-4 by Douglas configuration engineer and accomplished artist R. G. Smith. (Harry Gann via Nick Williams Collection)

The mock-up XF4D-1 presented to BuAer for preliminary inspection on 26 February 1949. (Boeing via Tommy Thomason Collection)

design—limited range, poor static thrust, and lack of reliability were a few of the challenges.

Manufacturers were reluctant to venture far from the shapes, structures, and systems that had worked so well in the past. But now the stakes were changing, supersonic flight was at hand, and the Air Force was developing a new generation of aircraft to take advantage of dramatic advances in aeronautical science. At the risk of being left behind, BuAer gradually embraced a few innovative designs.

George A. Spangenberg, former Director of Evaluation for BuAer described the Navy's caution during this period, "The Navy had a reputation at the time for building only conventional airplanes. Most of the industry thought we were well behind the Air Corps and unwilling to take aerodynamic risks. We had no tailless airplanes. In fact a lot of them thought the Navy was still in the biplane days, and Walter Diehl, our number one aerodynamicist, would never approve a tailless design; but he did."[4] Spangenberg was referring to the radical Vought F7U-1 Cutlass, which promised a huge leap in performance over the first generation of carrier jets. It was during this period of experimentation that Heinemann was invited by CDR (later ADM) A. B. Metsger, Head of the Fighter Branch of BuAer to discuss a possible role for D-571 in the Navy's defense planning.

Heinemann remembered CDR Metsger in his memoir, "A. B., as he was fondly called, and I had become close friends stemming from previous dealings that began with the XTB2D. A veteran carrier pilot with an extensive aeronautical engineering background, he was an ideal choice for the job. He wasn't initially sold on the delta-wing concept but his assistant, CDR Turner Caldwell of Skystreak fame, was. And it was Caldwell who soon convinced his boss of the fighter's virtues."[5]

Heinemann's designs were held in high regard by BuAer, but D-571 was so radical that it was a hard sell to the Navy "brass." Aerodynamicist A.M.O. Smith recalled that during one of his presentations some BuAer officers voiced concern about the stability of the tailless design. Smith fabricated some crude cardboard models and held an impromptu symposium on delta-wing aerodynamics, finally convincing the doubters of the design's inherent stability.[6]

Heinemann said, "It is difficult to believe nowadays, but in those postwar years an atmosphere of skepticism surrounded many aviation developments, especially jet-propelled machines. A good number of officials in BuAer, for example, believed that a delta-wing plane like ours would flip end over end in flight. I wasn't worried about this, but one day we did conduct a rudimentary experiment in the El Segundo plant. We had several balsa-wood models with wingspans of about two feet,

A Douglas technician shows boarding steps in the mock-up XF4D-1. This feature was not retained in the flying prototypes or in the production aircraft. (Boeing via Tommy Thomason Collection)

The mock-up provided for a retractable ladder that was stowed in the left wing root. This feature was deleted from the flying prototypes. (Boeing via Tommy Thomason Collection)

which varied in their delta-wing shapes. They ranged from a pure flying wing to the Model 571.

"'Let's fly 'em,' I said to Leo and Amo. We enlisted some engineers, who rigged a net to catch the birds, and a photographer to record the event on movie film. Then, after the day-shift had left the plant and air currents were stabilized, we trooped up to the rafters. This was not a Marx Brothers production, but an observer or two down below may have thought so. Grown men at the top of a building with model airplanes they are about to toss into a net? But we were really professionals just trying to get height and still air to conduct a legitimate experiment.

"We launched each model straight out, applying a slight downward pressure on the trailing edge of the wing. The pure flying-wing model flipped over, and out of control, pretty much end-over-end all the way down. But the others, all of which had shapes similar to that of the Skyray, flew beautifully. The D-571 looked very stable. I must be quick to add that, in the case of the D-571, the center of gravity was well forward, much more so than that of the flying wing."[7]

After much deliberation CDR Metsger described a mission that was perfectly suited to the Model D-571, "...we want a plane that can climb to 40,000 feet within five minutes or better, and intercept high-flying Soviet bombers before they can get to a target."[8] It was noted by author Bill Gunston that Metsger's specification was historically significant. It was the first formulated by any Navy in response to the threat of a jet bomber carrying nuclear weapons. The Douglas design, according to Gunston, could be considered the first naval interceptor—as distinct from fighter—in history.[9] Metsger's specification became known as the "Ten Minute Killer Mission."

In June 1947 the BuAer issued a contract to Douglas for the preliminary investigation and engineering development up to and including the mock-up of an aircraft based on the Model D-571 design. The mock-up inspection was tentatively scheduled for June 1948 and work on Model D-571 proceeded in earnest. Heinemann enlisted the services of scientists from the National Advisory Committee for Aerodynamics (NACA) to find the optimum shape for their interceptor. Kermit E. Van Every, now Heinemann's chief aerodynamicist for the project, refined the planform until the final configuration was quite unique. The wingtips became rounded and the trailing edge of the wing became swept. The design could no longer be considered a true delta, but rather a low-aspect-ratio swept flying wing with a blended fuselage. The control surfaces consisted of full-span *elevons*, which provided both pitch and roll authority. Inboard of the elevons were two additional control surfaces used for pitch trim. These trimmers were very important in the transonic region in overcoming "Mach tuck," an extreme nose-heavy condition caused by supersonic airflow. The proposed wingspan was only 33 feet 6 inches, but the wing area was an enormous 557 square feet giving the proposed interceptor the lowest wing loading of any

The mock-up tail bumper and tail hook. While this feature appears to be employed on the flying prototypes, it was substantially modified prior to the carrier trials. (Boeing via Tommy Thomason Collection)

A Westinghouse J40 mock-up being trial fit in the XF4D-1 airframe engine bay. (Boeing via Tommy Thomason Collection)

The rocket pod installation under the right wing. (Boeing via Tommy Thomason Collection)

A major change requested by the mock-up committee was the reshaping of the nose to improve visibility during carrier approaches. This is the revised nose shape displayed to the mock-up committee 10 days after the initial inspection. (Boeing via Tommy Thomason Collection)

One month later Douglas was still working on the nose contours. Here the mock-up is elevated to the expected carrier approach angle of more than 18 degrees to check forward visibility from the cockpit. (Boeing via Tommy Thomason Collection)

The original instrument panel layout shown to the mock-up committee on 26 February 1949. (Boeing via Tommy Thomason Collection)

The revised instrument panel on 7 April 1949, which has been changed in response to the mock-up committee recommendations. (Boeing via Tommy Thomason Collection)

contemporary fighter. The wing's gracefully rounded tip and swept trailing edge gave the appearance of a manta ray. In keeping with the Douglas custom of using the prefix *Sky* for many of their products (Skyknight, Skyraider, Skystreak, Skywarrior, etc.), the name *Skyray* was selected for the new design.

As Heinemann's design team refined the interceptor configuration, the Navy inserted numerous new design requirements. The mock-up, therefore, was not ready for review until March 1949. Some of the major recommendations from the mock-up committee included: 1) changing the contours of the nose to provide better visibility for carrier approaches; 2) changing the wing fold break and using power wing folding to provide more deck storage; and 3) increasing the engine bay diameter from 52 to 58 inches to accommodate the increased diameter of the afterburner of the proposed engine.[10]

The mock-up was approved in April 1949 and a contract was awarded in May for two flying prototypes and one static test article. The contract provided that the prototypes were to carry four 20mm cannon mounted in the wings. There would be six external store points to carry a combination of 4,000 pounds of ordnance. The cockpit would be pressurized to 5 pounds per square inch to provide pilot efficiency at high altitude, and it would have an ejection seat for emergency egress. The controls would be hydraulically boosted with manual reversion, and the tricycle landing gear would be supplemented with a small retractable tail wheel to prevent damage to the tail cone from the nose-high landing attitude. The contract also specified that the XF4D was to be powered by a new turbojet that was being developed by the Navy's primary jet engine vendor, Westinghouse. This new engine, designated the J40, promised a dramatic level of thrust in its afterburning form—more than 10,000 pounds, an unheard of amount of power for the time.

But the most important contractual requirement was maximum takeoff weight. Heinemann had become obsessed with weight ever since he ran afoul of the Navy's weight limits in 1944 during early SB2D production. George Spangenberg said, "I was able to show Leo Devlin, Heinemann's very capable assistant, that many of Douglas' design items were much heavier than those being produced by other Navy contractors. As example, I remember, was that the SB2D rudder pedals were nearly twice as heavy as those on Grumman and Vought designs. Douglas took the message to heart and started what became probably the most effective design weight control effort in the industry."[11] The Skyray contract provided for a takeoff weight limit of 16,821 pounds and

The first prototype at rollout on 9 October 1950. The markings are incomplete and the Bureau Number (BuNo) is incorrectly painted as 184586 rather than 124586. (Boeing via Nick Williams Collection)

Various views of the first prototype XF4D-1, BuNo 124586, in December 1950 just prior to the maiden flight. The aircraft is painted in overall Sea Blue with the markings that it would wear throughout early Phase I test flights. The nose stripe and the Skyray name were painted red with a white outline. The ground stance was level with the runway, and the nose gear door was one piece, which remained open when the gear was extended. All of these features would change on the production aircraft. (Boeing via Nick Williams Collection)

Skyray
NAVY

124586
NAVY

124586
NAVY

Heinemann was determined not to exceed this weight because the interceptor's climb performance depended on it. As insurance Heinemann chose to skin the airframe with very thin gauge aluminum (1/32 inch) stiffened underneath with a layer of stamped stringers to keep the structure as light as possible.

The three prototypes (two flying articles and one static test) were constructed under heavy security in an isolated corner of El Segundo's Building 1. Heinemann, who had misgivings about using the unproven Westinghouse J40 in a brand-new airframe, convinced BuAer to permit the installation of the reliable Allison J35-A-17A for the initial test flights. The Allison produced only 5,000 pounds of thrust (lbf), significantly less than the thrust expected from the Westinghouse, but it was deemed sufficient to fly the XF4D-1s in the early phases of the test program. Heinemann designed the engine mounts to accommodate the small J35 and allow for the larger J40 when it became available. A new tail cone was fabricated for the small diameter of the J35 tailpipe, the first of many tail cones that would be fitted to the Skyray. Structural tests on the static airframe showed the need for major revisions to the wing.

By October 1950 the first prototype, which was assigned Bureau Number (BuNo) 124586, was shop complete. BuNo 586 was painted in the typical Sea Blue scheme of that period, which was particularly handsome on the contours of the blended wing airframe. A long pitot boom for gathering accurate test data was fitted to the nose cone, and the nose was decorated with a red stripe outlined in white with the name *Skyray* painted above it. The elevons had no counterbalances on their tips, and the engine air intakes had a sharp triangular shape with no boundary layer ramp. The rudder was a single surface spanning the full height of the fin. Early photographs of 586 show no evidence of leading-edge slats on the wings. They may have been fixed in the retracted position for the initial test flights. Many of these features would change during the Skyray's development, which would unfold over a seven-year period.

After engine run-ups the XF4D-1 was prepared for trailer transportation to the Douglas Testing Division Facility at Edwards Air Force Base (AFB), Muroc, California. There, company pilots would explore the fundamental airworthiness of the bat wing interceptor in Phase I testing. The flight envelope would be expanded gradually to greater airspeeds and higher altitudes until the design limits of the aircraft were reached. Along the way necessary modifications would be employed to ensure that the Navy was getting the airplane it bargained for. When certain test milestones were achieved the Navy would take possession of the XF4Ds and fly them with their own pilots in Naval Preliminary Evaluations (NPEs), to verify Douglas' progress. Most of the flights would occur at Edwards, but on occasion the aircraft would be ferried to the Naval Air Test Center (NATC) at Patuxent River, Maryland, for specific tests such as carrier suitability trials or armament test.

A direct rear view of the first XF4D-1 showing the small-diameter tail cone that was fabricated for the Allison J35-A-17A interim engine. (Boeing via Nick Williams Collection)

America's Other Delta

It is interesting to examine the development of another delta-wing aircraft that evolved from the Lippisch data acquired by the Department of Defense Technical Mission in 1945. The Air Force F-102A Delta Dagger was designed, tested, and reached operational service at approximately the same time as the Navy F4D-1 Skyray. Both aircraft employed delta-wing configurations and both were powered by the Pratt & Whitney J57 turbojet engine.

In mid 1945 the Army Air Forces (AAF) issued requests for proposals for a point-defense interceptor that was capable of supersonic flight (defined then as 700 mph), and able to reach 50,000 feet in 4 minutes. Proposals were received from Bell, Douglas, North American, and Northrop, but the winning proposal came from Consolidated-Vultee, later known as Convair. Convair's initial design was an aircraft with a highly tapered, low-aspect-ratio, 35-degree swept wing and a butterfly "V" tail. The most exotic part of the proposal was the powerplant. At that time turbojet power was insufficient to achieve the Army's performance standards, so a radical—almost science fictional—powerplant was proposed. It consisted of a 3,700-lbf ducted rocket/ramjet with four external 1,200-lbf rockets. In addition a 1,560-lbf Westinghouse J30 axial-flow turbo jet was to provide thrust for landing, and generate electrical and hydraulic power throughout the flight. The Army designated this winning proposal the XP-92, and design work began in May 1946. As yet, Capt. Yeager and the Bell X-1 had not broken the sound barrier, but the XP-92 was contractually required to be supersonic.

Wind tunnel testing of a 1/20-scale model revealed significant problems with Convair's initial design. The wingtips of the highly tapered swept wing stalled at 5 degrees angle-of-attack, and poor lateral control was evident. Convair chose to abandon the tapered-wing butterfly-tail design and investigate a delta configuration. The delta offered several benefits: Its low-aspect ratio allowed the use of a very thin wing, which minimized drag; the low-aspect ratio provided rigid structure; and the large internal volume provided room for ample fuel capacity. Furthermore, Convair designers recognized the stable, stall-resistant flying qualities of a delta.

It was realized from the start that the XP-92 proposal presented daunting technological challenges. Not only was the propulsion system extremely complex, but also aerodynamics of delta wings were unknown at that time. In September 1946 Convair convinced the Army that it was impractical to develop an advanced delta-winged airframe with a novel ducted rocket/ramjet propulsion system simultaneously. The Army concurred and granted approval to separate the aerodynamic issues from the propulsion issues. As a first step Convair would build a delta-winged aerodynamic demonstrator using a proven turbojet engine. This flying mock-up was designated the XP-92A.

Work continued on the ducted rocket/ramjet version, but when the full-scale mock-up was inspected on 17 April 1948 the Air Materiel Command report concluded, "The airplane in its present configuration is highly impractical for any use other than research purposes." The exotic propulsion portion of the contract

Artist rendering of the original XP-92 concept. This configuration was abandoned when wind tunnel data revealed serious lateral control and tip stall problems. (Convair via the San Diego Aerospace Museum Archives)

Artist rendering of the revised configuration for the XP-92 using a large barrel-like fuselage as a duct for the radical propulsion system and delta flying surfaces. Both the propulsion system and the aerodynamics of this concept were extremely adventurous for the late 1940s. (Convair via the San Diego Aerospace Museum Archives)

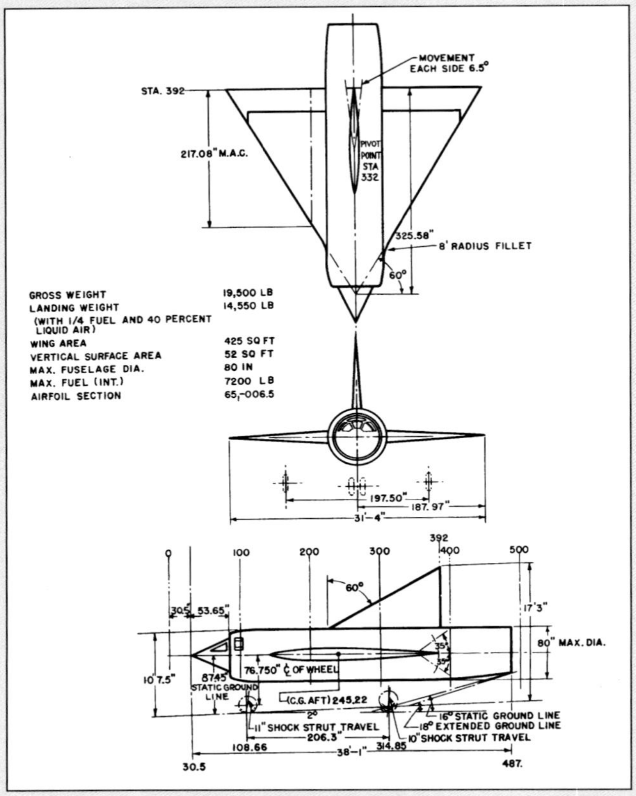

was canceled; however, the delta-winged jet powered aerodynamic demonstrator, XP-92A, survived.[12]

Convair assigned the model number 7002 to the XP-92A (later XF-92A) research airplane. It was to be built in the shortest period of time using available materials and parts to the maximum extent. The J33 engine from a Lockheed P-80 would power the experimental delta. The main landing gear was cannibalized from an FJ-1 Fury, the nose gear from a P-63, the brakes from a P-80, the ejection seat from the YP-81, along with scores of components from other existing aircraft.[13]

The airframe consisted of a 42-foot 6-inch-long, circular cross-sectioned fuselage with a clear frameless bubble canopy. The engine air intake was a circular opening in the nose. The wings, which were triangular in planform, had a span of 31 feet 4 inches and were attached at the horizontal centerline of the fuselage. The vertical stabilizer was also triangular with the rudder

General arrangement drawing of the delta-winged XP-92 proposal showing the triangular shape of its flying surfaces. This design was a highly adventurous attempt to meet the performance requirements that were considered nearly impossible for the time. (Convair via the San Diego Aerospace Museum Archives)

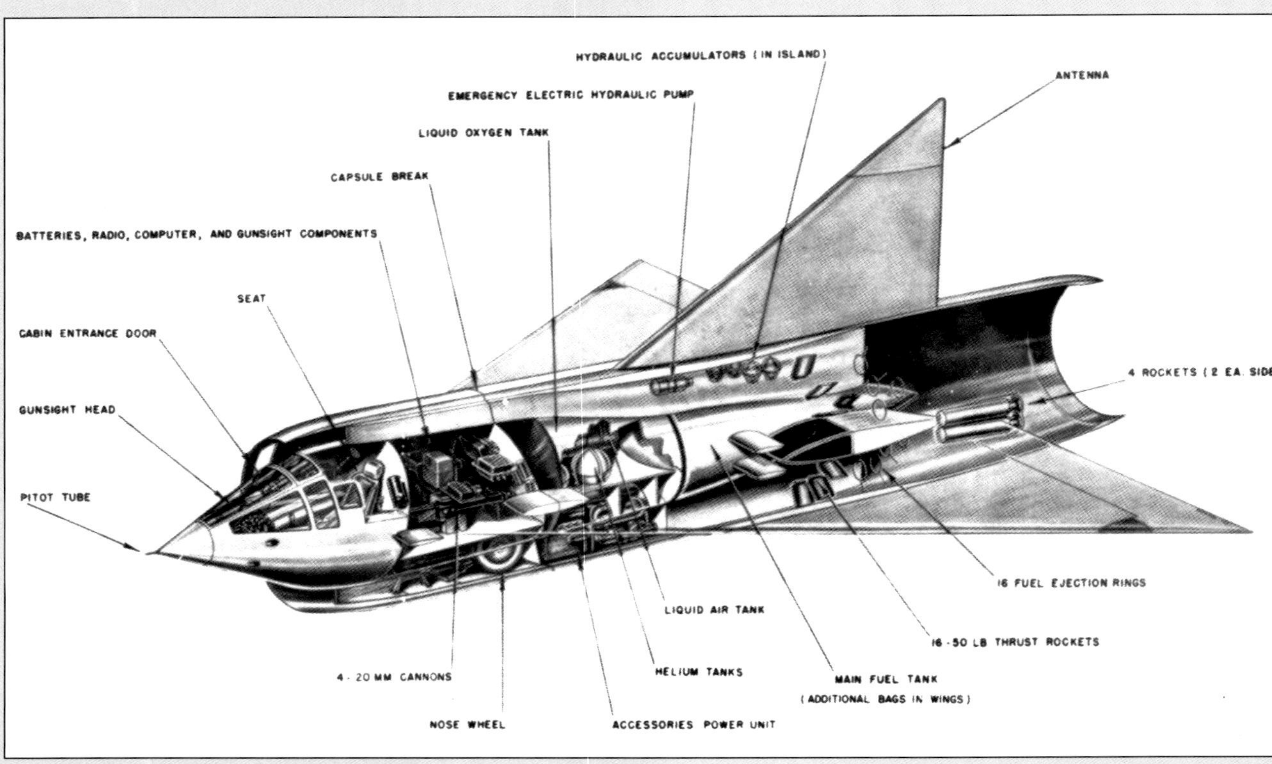

Cutaway drawing of the XP-92 proposal showing the complex ducted rocket propulsion system and the spiked cockpit arrangement. During the mock-up process Convair convinced the Air Force that the propulsion system and the delta-wing aerodynamics should be explored separately. (Convair via the San Diego Aerospace Museum Archives)

The XP-92 full-scale mock-up during Air Force inspection in April 1948, approximately one year prior to the Navy's inspection of the XF4D mock-up. The Air Force recognized during this inspection that the XP-92 propulsion system was impractical and canceled that part of the contract. (Convair via the San Diego Aerospace Museum Archives)

The Air Force awarded Convair a contract to study delta-wing characteristics by building an aerodynamic demonstrator. This aircraft was designated the Model 7002 by Convair and became the XF-92A. This is the full-sized XF-92A undergoing wind tunnel testing at NACA Ames. At this time the airframe had a clear bubble canopy. (NASA via Mike Machat)

trailing edge aligned with the wing trailing edge. The controls consisted of full span elevons, which acted in opposition for roll control, and in unison for pitch control. Unlike the XF4D-1 the XF-92A had no separate control surfaces for pitch trim.

Construction of Model 7002 was completed at the Convair San Diego plant on 31 October 1947. The airframe was unpainted natural aluminum and it wore a large national insignia on either side of the fuselage below the canopy. It was shipped to the NACA Ames Aeronautical Laboratory without an engine for full-scale wind tunnel tests to obtain data on the delta configuration. It was returned to Convair, San Diego, on 12 January 1948 and an Allison J33-A-21 engine was installed for test flight. On 26 March 1948 the aircraft was trucked to Muroc, California, for flight testing.

The XF-92A at the Convair facility in San Diego, California, during initial engine runs. At this time the aircraft was natural aluminum and carried sparse markings. The original clear canopy had been replaced with a framed canopy. (Convair via the San Diego Aerospace Museum Archives)

The XF-92A on the Convair seaplane ramp in San Diego, California, preparing for its first flight. This aircraft was built in a limited amount of time and used many components from other aircraft such as landing gear and engine systems. (Convair via the San Diego Aerospace Museum Archives)

Chapter Two

XF4D-1 Flight Test Program

"Climbing like a homesick Angel" was the catchphrase for dramatic maximum-performance takeoffs like this one. Here we see an early-production YF4D-1 Skyray taking to the skies for another flight test mission from Runway 06 at Edwards AFB South Base complex. (Nick Williams Collection)

Testing any new airplane is a challenge. Testing a new jet-powered, delta-wing, tailless prototype interceptor in the early 1950s takes this concept to the limit. Only the best need apply.

Mr. Peyton Takes the Ride of His Life

Larry Peyton, a Douglas test pilot from the Santa Monica Division, was selected to make the initial XF4D-1 test flight. Peyton was fundamentally a transport aircraft test pilot with limited experience in jets, even less in single-engine fighters, and certainly no experience in tailless deltas.

There were many unknowns about the radical airplane Peyton was expected to fly, but the greatest uncertainty involved the flight control system. Douglas engineers decided to make the first flight using the manual flight control configuration and bypass the complicated hydraulic power-assist system. The manual control system required lengthening the control stick for more mechanical advantage, and limiting the surface deflection

During the initial test flights both XF4D-1s wore an overall Sea Blue color scheme with a red nose stripe outlined in white. The tail cone was natural metal. All markings and stenciling were white. (Boeing via Nick Williams Collection)

to half of the normal amount. In the manual mode with the elevon deflection limited, it was suspected that Peyton would not have sufficient pitch authority to rotate; therefore, the pitch trimmers would have to be used. The amount of nose-up trim necessary to achieve rotation was unknown so the Douglas engineers advised Peyton to use full trimmer deflection for the first takeoff. Also, in keeping with normal practice on first flights, it was decided to keep the landing gear extended—a decision that would prove to be very fortunate for Peyton.

At 10:20 on the morning of 25 January 1951, at Muroc, California, Larry Peyton was about to have the ride of his life. He spooled the little Allison turbojet to 95-percent RPM and accelerated to 78 knots, allowing him to lift the nose wheel from the lake bed. At 121 knots the XF4D-1 became airborne, but as the Skyray continued to accelerate it became apparent that the nose-up trim was excessive. Peyton forced the lengthened control stick forward, yet the nose continued to rise. He realized that the nose-up force from the trimmers was excessive, but when he moved the switch, the trimmers moved too rapidly causing a sudden drop of the nose. He pulled back on the stick but couldn't get the nose to level flight. Another application of *up* on the overly sensitive trimmer switch caused the Skyray to flare and settle to the runway. Nevertheless, Peyton had the presence of mind to stay with the airplane and work through the oscillations.

When the Skyray became airborne again, Peyton reduced the power slightly and the pitch excursions became controllable. Peyton was able to achieve a stabilized climb at 140 knots. He leveled off at 10,000 feet and began checking the handling qualities. He found the Skyray to be very solid in roll, but highly sensitive to rudder inputs. The rudder was providing a significant roll component in addition to yaw. Peyton also discovered that the manual control system exhibited excessive slop. He could move the stick 6 inches fore and aft, or 4½ inches to either side with no control effect.

After investigating the Skyray's power-off characteristics, Peyton attempted a landing approach from 2,000 feet. At the runway threshold he reduced power and began to flare, but the airplane settled rapidly. He applied full nose-up trim and the Skyray ballooned. Again he forced the stick fully forward and touched down on the main gear and tail wheel. In Peyton's flight report he criticized the control system slop, and the excessively speedy pitch trimmers. Peyton never flew the Skyray again.[1]

The two XF4D-1s parked together on the ramp at Edwards Air Force Base during early Phase I testing. The second prototype, BuNo 587, has a ventral fin attached to its tail cone. (Boeing via Nick Williams Collection)

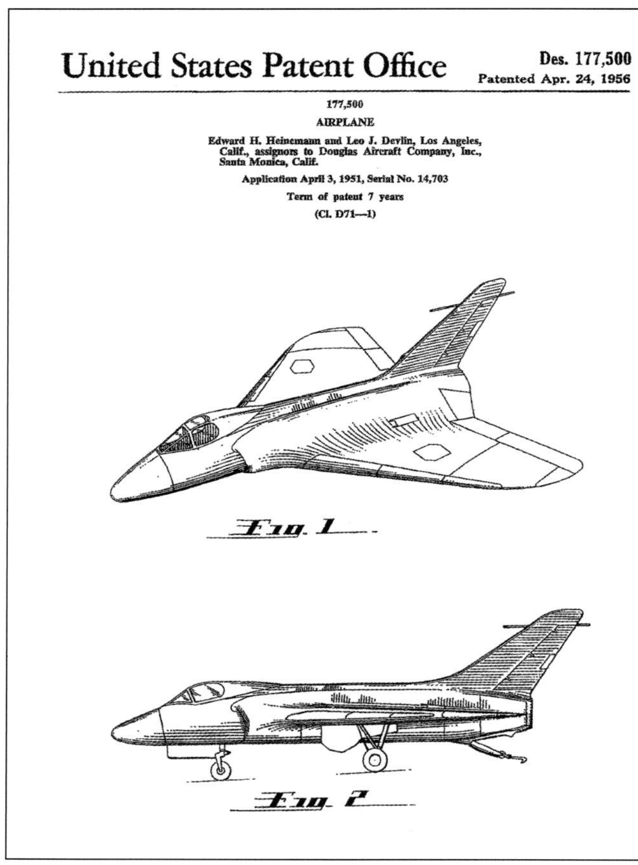
United States Patent Office

Des. 177,500
Patented Apr. 24, 1956

177,500
AIRPLANE
Edward H. Heinemann and Leo J. Devlin, Los Angeles, Calif., assignors to Douglas Aircraft Company, Inc., Santa Monica, Calif.
Application April 3, 1951, Serial No. 14,703
Term of patent 7 years
(Cl. D71—1)

Fig. 1

Fig. 2

Drawings submitted by Ed Heinemann and Leo Devlin as part of their patent application shortly after the initial test flights in 1951. The patent was granted in 1956. (Mark Nankivil Collection)

Heinemann recalled the first Skyray flight by saying, "The Ford's first half-hour in the sky was characterized by some control difficulties, not an awfully unusual occurrence in this business, but overall we knew we could work out the bugs and that essentially we had the makings of a fine fighter."[2] Two weeks after Peyton's white-knuckle flight BuAer issued a Letter of Intent, NOa(s) 51-637, for 12 production aircraft.

Counterbalances were added to the elevon tips to ease the stick loads in the event of a hydraulic failure. (Author's Collection)

Tail buzz at high speeds was an ongoing problem with the XF4D-1s. This illustrates some of the solutions that Douglas engineers tried. The problem was never completely resolved. (Boeing via Nick Williams Collection)

Bob Rahn Tests the Skyray

Several months after Peyton's difficult maiden flight a second Douglas test pilot, Russ Thaw, was assigned to continue the Skyray's test program. He completed 25 flights expanding the flight envelope and exploring successive fixes. In July the second XF4D, 587, arrived at Edwards to join the test program. Thaw took 587 aloft for its first flight on 27 July 1951. This Skyray differed from the first by having horn-balanced elevons (the elevon tips were modified to carry a counterweight forward of the hinge line to offset the weight of the surface and also provide an aerodynamic balance when deflected to reduce flight loads). Thaw found that this change made a dramatic improvement in the control system.

Meanwhile, Bob Rahn, a highly regarded experimental test pilot, became available to enter the XF4D program. Rahn made his first Skyray flight in October 1951 and was immediately impressed. He surmised that the aircraft had the potential to become an exceptional fighter. On his third flight, Rahn flew on the manual control system that Peyton used on the first flight, and it was a nightmare! The control forces were overwhelming and the airplane was difficult to handle. Rahn insisted that the Skyray be ferried to El Segundo for the installation of an emergency electrically driven hydraulic pump and a yaw damper. In January 1952 the Skyray was returned to flight status and Rahn again evaluated the manual system. By now the system was far more manageable, "Under ideal conditions I could actually land the Skyray on manual without touching the stick by using the rudder for roll and the trimmers for pitch control," said Rahn.[3] Nevertheless, Douglas decided to design a redundant hydraulic system to avoid the problems with the manual reversion system.

Transonic-speed flight explorations were begun even though the Skyray was still powered with the 5,000-lbf

BuNo 587 wearing its new white color scheme and powered by a Westinghouse J40-WE-6 against the California desert sky. (Boeing via Nick Williams Collection)

BuNo 587 displaying the new tail cone needed to accommodate the Westinghouse J40-WE-6 engine. (U.S. Navy via Nick Williams Collection)

BuNo 586 after the application of its white paint scheme. The clamshell canopy has been installed, and only the inner portions of the slats are extended. (U.S. Navy via Nick Williams Collection)

Allison J35. These high-speed tests were conducted in a shallow dive from 35,000 feet. Rahn noticed that a moderately severe tuck-under was encountered at Mach .91 requiring large pitch trimmer deflections. Also, rudder buzz and tail buffeting was pronounced. Fourteen tail cone modifications were tried to eliminate the buffet but none was completely successful.

By February 1952, more than a year after the maiden flight, Skyray 587 was finally fitted with a non-afterburning version of the Westinghouse J40 engine. This was not the engine specified in the Navy contract, but it provided an additional 2,000 lbf over the interim Allison J35. The performance of 587 was measurably improved; the re-engined XF4D could climb to altitude faster and reach high transonic speeds in a nearly level attitude. The appearance of both XF4D-1s changed in mid 1952. The Sea Blue aircraft were hard to spot in the California skies so they were repainted in a glossy white scheme. The tail cones, which were undergoing frequent modification, were unpainted natural metal. The nose retained the stripe and the name *Skyray* in red.

Spin testing was the next order of business for the Skyray in its new white scheme. The re-engined 587 was designated to fly the spin tests with Bob Rahn at the controls. No delta-winged airplane had ever been spun before, and Rahn planned to approach the tests with unusual caution. The initial spins were conducted with a forward center-of-gravity making the aircraft more inherently stable. In this state, the Skyray recovered using the conventional technique of opposite rudder and neutral stick. Then the center of gravity was moved aft, and all hell broke loose. The Skyray began to oscillate in the spin; it sometimes reversed the direction of spin rotation and the number of turns necessary for recovery increased. Worse, in a left-hand spin Rahn discovered that the Skyray would enter an inverted spin—a very disorienting and difficult condition. Rahn remarked, "Such spins, as every aviator knows, are hazardous to one's health." A spin chute was automatically deployed at 10,000 feet allowing the tumbling Skyray to stabilize. Rahn jettisoned the chute at 4,000 feet and regained level flight. The spin had started at 36,000 feet; it required 32,000 feet to recover.

Gradually, Rahn and the Douglas engineers developed an effective spin recovery technique dramatically different from conventional technique. The turn-and-bank indicator was used to determine the direction of the spin. The stick was held full aft to keep the Skyray in a positive stalled attitude, preventing an unplanned inverted spin. The ailerons became the primary recovery control and were applied in the direction of the spin. The rudder was ineffective since the stalled wing blanked the airflow over the tail.[4]

On 25 April 1952 the Navy, pleased with the Skyray's progress, placed an order for 230 production aircraft with Letter of Intent, NOa(s) 52-977. Douglas had not yet demonstrated supersonic flight, but the Navy insisted on evaluating the airplane in its current state of development. Douglas was able to delay the evaluation for several months since 587 was being used for spin tests and 586 was having its non-afterburning J40 engine installed.

On 15 July 1952 an evaluation team of Marine Maj. Marion Carl, CDR Bob Elder, and LCDR Bob Clark from NATC Patuxent River arrived at Edwards Air Force Base to fly 586. A total of 7½ hours were flown during 10 flights. The evaluation team flew simulated carrier approaches and was generally satisfied with the Skyray's performance, but the new Westinghouse engine was a disappointment. During ground measurements Douglas engineers were only able to record 6,000 pounds of installed static thrust, possibly due to the 2,300-foot elevation at Edwards. This was nearly 45-percent-less thrust than the Navy expected with the afterburning Westinghouse J40. The evaluation was conducted with a maximum speed restriction of Mach 0.95. Nevertheless,

BuNo 586 after receiving its Westinghouse J40-WE-6 engine. The tail cone is slightly different than the tail cone fitted to BuNo 587. (U.S. Navy via Nick Williams Collection)

BuNo 586 making a field arrested landing at Edwards Air Force Base during the contractor demonstrations. Douglas was required to show that the aircraft was structurally capable of flying the carrier suitability trials safely. (U.S. Air Force via Nick Williams Collection)

Four views of BuNo 586 during the contractor demonstrations prior to the Navy's carrier suitability trials. The XF4D-1 has movie cameras mounted on the left wingtip, belly, and fin. 150-gallon fuel tanks are mounted on each pylon. The launch bridle is secured to the fuselage for test purposes. (U.S. Navy via Nick Williams Collection)

the team was highly impressed with the Skyray's agility at high speed. They reported, "At 40,000 feet in wind-up turns to the left, steady accelerations without airframe buffet at about 3.4 g at M-0.90, 3.2 g at M-0.80, and 3.0 g at M-0.70 were obtained. The F-86E chase plane was lost immediately on such maneuvers."[5] However it was noted that the limited thrust from the J40 caused a considerable loss of altitude.

Another thrust-related problem was the time-to-climb performance. From a standing start it took the XF4D-1 more than 18 minutes to reach 40,000 feet. The F-86E could maintain formation with the climbing Skyray at 97-percent RPM, but the Skyray could easily outrun the F-86E in level-speed runs. The Preliminary Evaluation summary stated, "The evaluation team was particularly impressed by the capabilities and potentialities of this airplane as being superior to anything flown to date. It is recommended that every effort be made to hold the aircraft at its present weight and provide it with sufficient power to realize the performance of which it is capable."[6]

Taking note of the Skyray's ability to maneuver at altitude, easily out-turning the Air Force chase planes that were assigned to the evaluation, Maj. Carl commented, "If we had this airplane now in Korea, I could just pop off the MiGs—one, two, three."[7] Numerous modifications were recommended as a result of the Preliminary Evaluation. Douglas chose to address many of these issues in the production aircraft, but some changes were made immediately. Most significantly, the control system of 586 was improved to the standard of 587 by the addition of counterbalanced elevon tips to relieve extreme stick forces in the event of a hydraulic power loss.

With the Naval Preliminary Evaluation (NPE) successfully concluded in the summer of 1952, Bob Rahn continued his investigation of high-speed flight in the Skyray. He was eager to take the aircraft supersonic since no delta-wing airplane had ever flown above Mach 1 before. The record is not clear which airplane was used, nor is the exact date certain, and when the historic milestone was achieved Rahn was unaware of it. A faulty Mach meter read .97 but all of the ground observers clearly heard the sonic boom. On the second dive Rahn got the proper instrument reading leaving no doubt that the Skyray had successfully been through the sound barrier and flown at supersonic speed.[8]

Carrier Suitability

Meanwhile, Douglas prepared 586 for one of the most important phases of the entire test program, carrier suitability trials. BuAer had recently implemented a requirement that the manufacturer demonstrate simulated carrier launches and arrestments on land

BuNo 586 on the deck of USS* Coral Sea *(CVA-43) in late October 1953 during the carrier suitability trials. (U.S. Navy via Nick Williams Collection)

before the Navy commenced the actual sea trials. This allowed the Navy to monitor the structural integrity of the airframe during high-stress operations and make corrections before the new airplane went to sea. The first catapult launches were made at Inyokern, California, on 6 May 1953 followed by six successful simulated carrier landings at Edwards.

In July 1953 Douglas felt confident enough to ferry Skyray 586 to the NATC, Patuxent River, Maryland, where company pilot J. O. Seay was assigned to complete the manufacturer's demonstrations in front of NATC engineers. The demonstrations were flown at gross weights ranging from 16,000 to 20,500 pounds. The launches were flown both clean and with full external stores mounted under the wings.

From 23 July to 31 July Seay made 13 field catapult launches and 12 field arrested landings. Aircraft 586 carried an instrumented launch bridle that measured catapult towing loads. This bridle was secured to the Skyray and did not fall away after the launch. The arresting hook on 586 was also instrumented to measure the loads imposed on each landing. Wingtip-, fin-, and belly-mounted cameras recorded the action from the airplane, while ground-mounted cameras captured the catapult and arresting cables.

The catapult at Patuxent River was an H4B with a 96-foot stroke. At full gross weight this catapult could accelerate the Skyray to only 80 knots. Fortunately, the runway provided another 3,000 feet for the Skyray to reach flying speed. The arresting gear was the Morest

BuNo 586 climbing after a launch from the USS Coral Sea ***(CVA-43) during the carrier suitability trials in October 1953.*** (U.S. Navy via Nick Williams Collection)

BuNo 586 being brought topside to the flight deck of USS Coral Sea ***(CVA-43) on a plane elevator in preparation for launch.*** (U.S. Navy via Rob Getz, www.steller-views.com)

BuNo 586 departing the deck of the USS Coral Sea ***(CVA-43) during the carrier suitability trials.*** (U.S. Navy via Nick Williams Collection)

type with two 1-inch-diameter cables spaced 40 feet apart. The cables had 293 feet of runout. The first arrested landing was a very-high-energy event. The Skyray carried 2,200 pounds of internal fuel and Seay flew the approach at a brisk 140 knots. The fuel-laden aircraft engaged the wire and pulled out all of the available cable. The instrumentation showed that 135 percent of the arresting hook load limit had been reached but a careful inspection confirmed that no structural damage had occurred. On subsequent flights the internal fuel was reduced by 500 to 1,000 pounds.

During the following 11 traps Seay intentionally demonstrated every poor landing possible, exceeding the design limits of the airplane on several occasions. The Skyray never broke, and the Navy was convinced that it was ready to go to sea.[9]

In late October 1953, the Navy assumed custody of 586 for the final carrier suitability tests. CDR Marshall V. Beebe, Director of Flight Test at the NATC, was assigned as project officer for the trials. LCDR James Verdin joined him. The tests were conducted in five phases: preliminary evaluations, launch tests, arrestment tests, barrier engagements, and deck handling.

During the preliminary evaluation several criticisms of the cockpit were noted. The headrest, the ejection-seat face curtain, and several switches were among the items noted. The XF4D-1s used two different canopy systems.

BuNo 586 making an arrested landing aboard the USS Coral Sea (CVA-43) during its carrier suitability trials. (U.S. Navy via Nick Williams Collection)

Bob Rahn prepares BuNo 587 with the afterburning J40 for the world absolute speed record. (Rob Getz, www.stellar-views.com)

A clamshell canopy, which opened upward, was used on 586, while a sliding canopy was used on 587. Verdin had recently flown the World Speed Record Trials in 587 and he preferred its sliding canopy. He stated that the clamshell was slow and difficult to open and close. Furthermore the clamshell canopy couldn't be opened during high-speed taxiing or in high-wind conditions. The sliding canopy could be opened to any intermediate position during taxiing, and it required no special locking operation.

The catapult phase consisted of 35 field launches from Patuxent River. Aircraft 586 was launched with various loads of internal fuel, with and without external fuel tanks, and with various rocket packages. The airplane was positioned off center and with the nose wheel misaligned to determine its tracking characteristics.

On 23 October, following the successful completion of the field launches at Pax River, the Skyray finally went to sea. CDR Beebe flew 586 to Chambers Field, Naval Air Station (NAS) Norfolk, where it was hoisted aboard the USS *Coral Sea* (CVA-43), and 14 deck launches were flown in various load configurations. Beebe remarked that his duties consisted of far more that just flying the test article. Each flight required careful planning with the Project Officer to determine what characteristic was to be tested and how it was to be measured. There was some telemetry on the XF4D-1 but the catapult crew took most of the measurements. After each flight a detailed flight report was composed, often taking 3 hours to complete.[10]

The arrestment phase consisted of 35 field landings at Pax River followed by 13 shipboard landings aboard the *Coral Sea*. The landings went smoothly. Beebe recalled that his first landing was probably the most flawless trap he ever made. "Bob Canaday was the Douglas rep on the *Coral Sea* and he couldn't get over the first landing. He said, 'You couldn't have centered it any better.' The nose wheel was right on the centerline of the deck, and there was no sway in the runout. Years later, whenever I ran into Bob, he would remind me of that landing. It was an easy landing airplane; I liked it."[11] The Skyray was deemed suitable for carrier operations; however, the final report recommended more than 20 changes be made to the production aircraft.[12]

World Speed Record

While 586 was being used by the Navy for carrier trials, 587 remained at Edwards and received the long-awaited afterburning J40-WE-8. The thrust of the new engine was impressive and the Navy's hopes for the interceptor were starting to be realized. Douglas officials became so enthused about the Skyray's performance with the afterburning Westinghouse engine that they submitted an application to the governing body for all world air records, the Fédération Aéronautique Internationale (FAI), for an attempt at the world absolute speed record over a 3-km course (1.86 miles). The Douglas application

Westinghouse finally made the afterburning J40-WE-8 available for the Skyray flight test program. Here XF4D-1 BuNo 587 is chained to the tarmac for a full power run-up. The installed static thrust of the J40 in afterburner was measured at 11,800 pounds. (Boeing via Nick Williams Collection)

was approved and the record attempt was scheduled for 26 September 1953.

The record attempt was to be flown by Navy LCDR James Verdin (Verdin would join CDR Marshall Beebe in the carrier suitability trials of 586 one month later). However, the task of preparing the XF4D-1 for the official flight fell to Douglas test pilot Bob Rahn who recalled, "Flight test personnel are usually considered pessimistic, but for this project we were extremely optimistic with the Skyray. We had 11 working days to install the engine, increase the speed by 100 mph, evaluate

Bob Rahn briefs LCDR Jim Verdin in the cockpit of XF4D-1 BuNo 587 in preparation for the world absolute speed record. (Rob Getz, www.stellar-views.com)

The tent set up in the California desert to shelter the equipment and personnel from the intense heat during the speed record trials. (Nick Williams Collection)

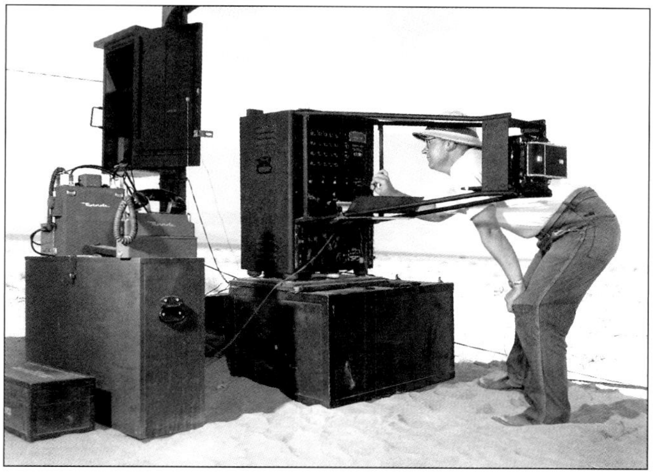

stability and control on the deck, prepare a flight plan, and become familiar with the course.

"I made the first unrestricted afterburner flight on the 20th, and to everyone's surprise, the engine worked flawlessly. After only three test hops with the afterburning engine, I flew the Skyray over the 3-km course at 100 feet of altitude.

"...I advanced the throttle beyond the détente [sic] in the throttle quadrant, igniting the afterburner. I felt a powerful push to my backside as the Skyray quickly accelerated to more than 700 mph.... The desert floor, a mere 100 feet below, became a blur and the airplane started buffeting from the rough air as it approached the speed of sound."[13]

In preparation for the speed trials, 587 was stripped of its tail hook and tail wheel. The tail hook fairing was sealed as was every other opening and protuberance on the airframe. The leading-edge slats were fixed in the retracted position, and the airplane was carefully waxed.

The Douglas and Navy strategy was to fly the record attempt under conditions that would allow the Skyray to fly at the highest possible true airspeed. High ambient air temperatures and a low-altitude course were desirable features for the speed run. It was decided to fly the record attempt 140 miles south of Edwards in the Salton Sea area of California. This site was actually below sea level and often experienced very high surface temperatures. The attempt required exceptional piloting skills to manage the aircraft at near supersonic speeds in the very dense and turbulent low altitude air. It also required exceptional judgment and reflexes to manage the afterburner, allowing the Skyray to accelerate to maximum speed just before entering the timed portion of the course, and then decelerate for an efficient turn-around without consuming too much fuel.

Rahn flew the afterburner-equipped Skyray for only 1 hour and 55 minutes before he turned it over to LCDR Verdin on 25

Officials from the Fédération Aéronautique Internationale (FAI) measuring the speed of BuNo 587 during the world record attempt on 3 October 1953. (Nick Williams Collection)

LCDR Jim Verdin prepares for the world absolute speed record attempt in the early morning of 3 October 1953. (Nick Williams Collection)

LCDR Jim Verdin returns to El Segundo in a corporate Super DC-3 after setting the world absolute speed record in XF4D-1 BuNo 587. (Nick Williams Collection)

LCDR Jim Verdin disembarks from XF4D-1 BuNo 587 after the speed record trials. (Nick Williams Collection)

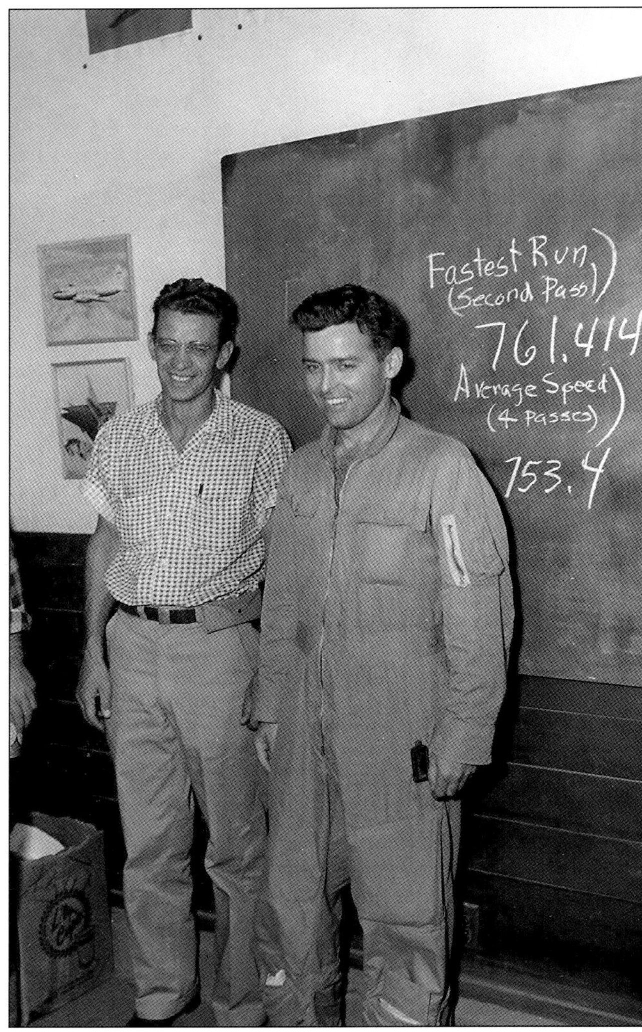

LCDR Jim Verdin receives the official results of his four speed-run passes. He has just been informed that he now holds the world absolute speed record. (Nick Williams Collection)

Douglas test pilot Bob Rahn congratulates LCDR Jim Verdin for setting the world absolute speed record. Rahn was responsible for preparing the XF4D-1 Skyray for the record attempt. (Nick Williams Collection)

Douglas publicity photos of LCDR Jim Verdin after the successful world speed record attempt. (Nick Williams Collection)

September. Verdin made several familiarization flights in preparation for the official record attempt that was scheduled for the next day; however, high winds and turbulence on 26 September forced cancellation of the attempt. Equipment repairs and additional weather problems postponed the flight until the following week. Finally, on 3 October Verdin was able to make four official passes through the timed course. The FAI measured the passes at 745.075, 761.414, 746.053, and 759.499 mph for an average of 753.4 mph—a new world speed record. This was the first time that a carrier aircraft held the record, but it was short lived.

A Douglas publicity photograph of the personnel responsible for the successful world speed record attempt. Test pilots Bob Rahn and LCDR Jim Verdin are in the middle of the back row. (Nick Williams Collection)

Douglas test pilot Bob Rahn in the cockpit of BuNo 587 prior to the 100-km speed trials. Rahn was the first civilian test pilot to set a world record in a military aircraft. (Nick Williams Collection)

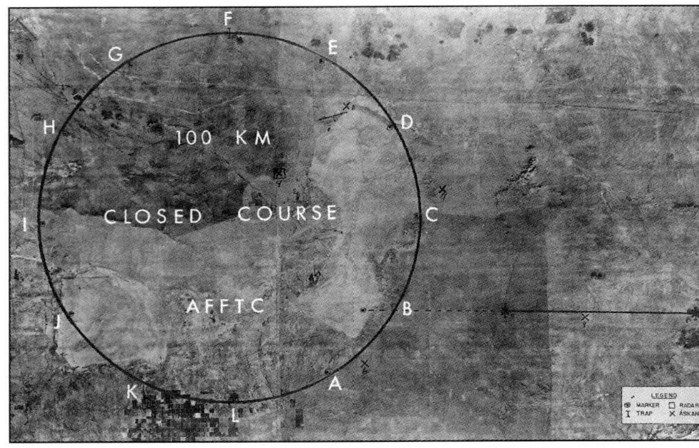

An Air Force photograph of the 100-km course in the Mojave Desert. (Nick Williams Collection)

Bob Rahn taxies to the active runway in BuNo 587 during the 100-km speed trials. The canopy on 587 slides rearward, while the canopy on BuNo 586 opened upward. Apparently a new nose cone has been installed for the 100-km flights, as the red stripe and the name Skyray is missing. (Nick Williams Collection)

Bob Rahn passes the finish line of the 100-km course setting the world record of 728.11 mph. The smoke is from burning tires to give Rahn a visual reference of the finish line. (Nick Williams Collection)

LCDR James Verdin (left) and Robert Rahn (center) accept medals for setting world speed records for 3-km and 100-km distances in the XF4D-1 Skyray. (Nick Williams Collection)

After the successful 100-km record attempt Rahn pulled the XF4D-1 into a steep climb and pitched into a hard left bank as he descended to the lake bed for landing. (Nick Williams Collection)

After successfully setting the world speed record, the Skyray was placed on public display at the Los Angeles International Airport in 1953. (Jim Sullivan via Nick Williams)

XF4D-1 BuNo 587 as it appeared during the world speed record attempts. (Nick Williams Collection)

Merely one month later, in November 1953, Lt. Col. Frank Everest Jr., set a new record in an Air Force YF-100. Flying over the same Salton Sea course Everest recorded a speed of 757 mph on the 3-km course. While FAI rules require a 1-percent increase to set a new record, and the YF-100 speed was not 1 percent greater than the Skyray, the Air Force took advantage of a little known technicality. The rules provide that the absolute speed record can be established by the fastest average flown over either a 3-km or 15-km straight course. Having failed at exceeding the Skyray's 3-km record by 1 percent a 15-km course was hurriedly prepared at the Salton Sea and Lt. Col. Everest flew it at 754.98 mph. While this performance was less than the earlier 3-km attempt, it was faster than the Skyray's record by nearly 2 mph, and it was not subject to the 1-percent rule. The official timer, Bertrand Rhine, submitted it to the FAI as the new world's absolute speed record and it was recognized.

There was quite a bit of animosity over North American's use of a "loophole" to capture the record. A Douglas spokesman was quoted in the 9 November 1953 issue of *Aviation Week*, "The F4D is still a faster airplane. They tried to top our F4D mark by the required 1 percent and couldn't do it. All previous records have been set on the 3-km course, including those of North American itself as well as the British and others. How come all of a sudden it's the 15-km course?"[14]

President Dwight Eisenhower presenting the highest award in American aviation to Skyray designer, Ed Heinemann at the Wright Brothers Memorial Dinner in Washington, D.C., on 18 December 1953. (Boeing)

The record-setting Skyray was scheduled to return to the flight test program, but Bob Rahn had a better idea. Since the airplane was configured for speed trials, he suggested to Donald Douglas, Jr., that they attempt to break the 100-km (62.1 miles) record. Douglas agreed and the Navy concurred on the condition that the record attempt would not delay the flight test program. Jim Verdin had been assigned to fly the carrier suitability trials in 586, so Bob Rahn was pleasantly surprised to learn that the Navy wanted him to fly the record attempt. He noted, "As far as was known, this was the first and only time that a civilian was allowed to set a world record in an airplane owned by the U.S. military."

The course was set up over the northern end of Antelope Valley, California, covering Rogers and Rosamond dry lake beds. The official run was made on 15 October 1953. Rahn kept the Skyray at 100 feet above the desert floor while he rounded the course pylons at nearly supersonic speed. "...I flew directly from one [pylon] to the next, rounding each turn in a tight 70-degree bank. After each turn, which encompassed a heading change of 30 degrees, I rolled the Skyray back into level flight for the next 15 seconds or so it took to reach the next pylon."

The flight lasted slightly more than 10 minutes from takeoff to landing and consumed more than 600 gallons of fuel. Quick calculations from the ground observers suggested that the closed-course speed record was easily set; however, the official timing mechanism had malfunctioned and the record attempt would not count. The flight was repeated the next day, and while this attempt was slightly slower than the earlier one, it was sufficient to establish the new world record of 728.11 mph. Rahn explained the reasons that the 100-km closed-course record was 25 mph slower than the 3-km straight course record, "There are four reasons: 1) higher altitude—2,600 feet mean sea level, which reduced the thrust; 2) lower temperature—22 degreesF, which resulted in lower true airspeed; 3) twelve 30-degree turns were made, which reduced speed on each turn; and 4) flying outside the pylons resulted in a greater distance than 100 km."[15]

Even though jet speeds continued to advance, the Skyray 100-km record stood for more than five years. It wasn't until 25 February 1959, that a Dassault Mirage III flying at 1,095 mph broke it. By that time the FAI rules had been changed to permit high-altitude attempts. As speeds increased, low-level flight became increasingly dangerous. The Skyray was probably the last aircraft to make a 100-km record at ground level.

World Altitude Record

The Navy was ecstatic over the Skyray's speed records and wanted to capture the world altitude record. Since the Skyray was designed as a fast-climbing interceptor, high-altitude flight was expected to be its strong point. A British Canberra bomber held the current record of 63,668 feet. It was learned that Skyray designer, Ed Heinemann, was to receive the Collier Trophy, an annual presentation for the highest achievement in American aviation. The trophy was to be presented by President Eisenhower at the Wright Brothers Memorial Dinner on 18 December 1953 in Washington, D.C. The Navy realized that it would be an enormous publicity coup to announce the altitude record at the dinner. In early December the Navy asked Douglas and Westinghouse to prepare 587 for the high-altitude attempt. LCDR James Verdin was assigned again as the record attempt pilot.

During the flight test program the Skyray had never been flown above 40,000 feet, so it fell to Bob Rahn to explore high-altitude flight with the afterburning Westinghouse J40-WE-8. During the low-altitude speed runs the engine had performed well but high altitude revealed numerous problems. Rahn recalled, "At the outset, balky afterburner controls developed in the J40 at high altitude. I had compressor stalls—explosions, really—which occurred at a rate of two per second. I experienced engine and duct rumbling. Then there was engine backfiring, complete with sheets of flame that extended no less than 10 feet out of the tailpipe. Plus, fuel starvation was causing afterburner blow out—a real craw in our collective throat.

"The main source of my anxiety had to do with complete engine failure and consequent loss of cabin pressurization. This would transform the pressure suit into a vise, with me inside it. The windshield would frost up, and I would have to go on gauges. Trouble was, the aircraft wasn't instrumented for blind, or instrument flying. I carried ice scrapers so that I could shave frost off the windshield.

"We were working on the problems right up to the morning of the day of the banquet, 18 December. I was in the air with the afterburner lit, climbing but didn't have smooth operation of the AB [afterburner] and was not climbing satisfactorily. After I landed, the crew made one small adjustment to the fuel control in the hopes that Verdin could try for the record in the afternoon. Up to this time I hadn't been able to get the Skyray much over 50,000 feet.

"Good as our intentions were, Jimmy could climb no farther than 25,000 feet as the fuel-control problems persisted. He was unable to even light the afterburner.

"At the banquet that night in Washington, D.C., no announcement of the new altitude record was made, but Ed Heinemann was awarded the prestigious Collier Trophy by President Eisenhower for development of one of America's first supersonic service fighters, the Skyray."[16]

LCDR Jim Verdin (left) and Bob Rahn wearing full pressure suits during the altitude record attempts. These suits were Air Force equipment. (Nick Williams Collection)

In this heavily retouched photo, ship 587 is shown in a maximum-performance 70-degree climb during the world altitude record attempts. Recurrent afterburner problems with the Westinghouse J40-WE-8 prevented the XF4D-1 from obtaining the record. (National Air and Space Museum via Nick Williams Collection)

The Skyray's speed records did not go unnoticed by the Air Force. In November 1953, the Air Force assigned four senior flight test officers to conduct a preliminary evaluation of the XF4D-1. The Air Force test pilots flew 11 flights totaling 5-1/2 hours.

During the Air Force evaluation the J40 afterburner blew out at 43,000 feet and relights were only possible below 25,000 feet. The Air Force report stated, "No special engine tests were flown to specifically investigate the quality of the J40-WE-8 engine; however, it was believed by the Air Force pilots that the engine was unsuitable and unsatisfactory in its present configuration. Numerous delays in the evaluation test program were caused by the engine and engine control deficiencies. The specific fuel consumption of the engine appeared to be high." The report recommended that the Navy install the Pratt & Whitney (P&W) J57 engine for improved reliability and increased performance.

Regardless of the noted engine faults, the report was highly enthusiastic about the airframe. The evaluation summary stated, "The Pilots' Evaluation of the XF4D-1 reveals that it is an exceptionally high performance airplane. It is one of the few airplanes that combines excellent low-speed handling characteristics with outstanding climb and high-speed ability. The short landing and take-off characteristics are comparable to our jet trainer types. The airplane is potentially adaptable to the Air Force air superiority, interceptor, and day fighter missions..."[17] The report concluded by recommending that the Air Force purchase the Skyray.

Notwithstanding the favorable test evaluation, the Air Force never seriously considered the procurement of the Skyray because it had already contracted for its own delta-wing interceptor, the F-102 Delta Dagger. Furthermore, Douglas could not meet Air Force deadlines because of its previous production commitments to the Navy.

BuAer rewarded Douglas with an order for 178 additional Skyrays. As yet, the first production aircraft had not been built, but the Navy had enormous expectations for the world-speed-record Skyray.

Second Career for the XF4D-1

The two prototype XF4Ds were remarkable airplanes. They represented a highly unconventional configuration that performed beyond expectation. Both flew at extreme altitudes and speeds for nearly five years and served brilliantly as proof-of-concept examples for the Navy's interceptor program. Both had gone through three engine changes, and set several world records along the way. But the XF4Ds were beginning to show their age, in fact, 587 had the name *Old Fireball* stenciled above its nose stripe. By the summer of 1954 Douglas was using the first block of production aircraft to complete the contractor's portion of the flight test program, so the Navy prepared to strike the worn prototypes. However, Virgil Weaver, Manager of General Electric's Flight Test Center at Edwards, identified the Skyray as

The J79 was shorter than the J40 that it replaced so it required an adaptor to mate the General Electric engine to the Skyray inlet ducts. (General Electric via Nick Williams Collection)

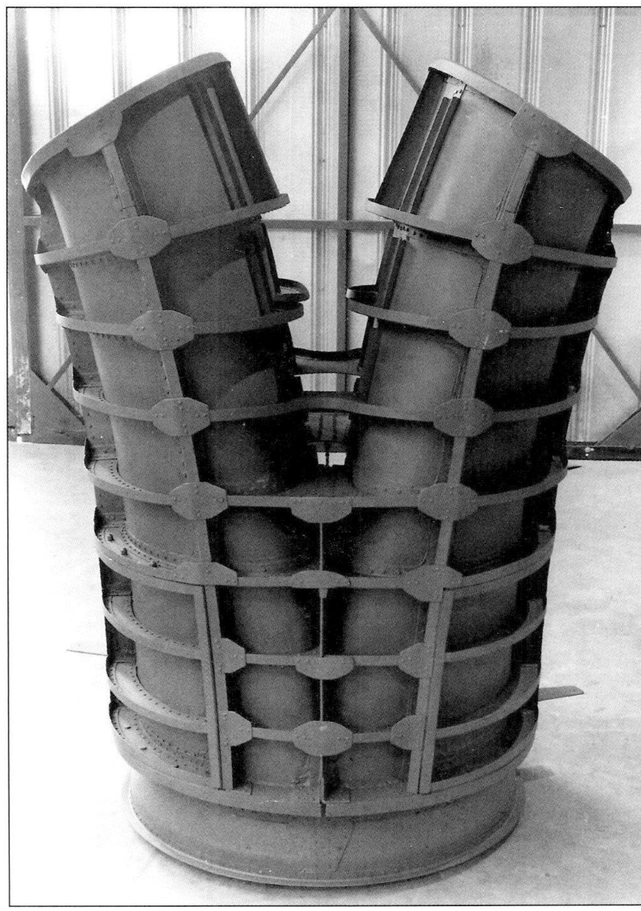

General Electric J79 installation in BuNo 587 with the afterburner nozzle open. (General Electric via Virgil Weaver, Nick Williams Collection)

General Electric J79 installation in BuNo 587 with the afterburner nozzle closed. (General Electric via Virgil Weaver, Nick Williams Collection)

BuNo 587 with the first airworthy General Electric J79 engine installed. This is a ground run in preparation for the first flight. (General Electric via Virgil Weaver, Nick Williams Collection)

BuNo 587, powered by the General Electric J79, retracts its landing gear at Edwards Air Force Base in February 1957. The J79 flights were usually flown clean with no external fuel tanks. (General Electric via Virgil Weaver, Nick Williams Collection)

the perfect test bed for the new J79 turbojet. Until that time new engines were test flown as secondary powerplants on large multi-engine airplanes such as B-45s or B-50s. Weaver realized that a new fighter engine should be test flown in a fighter environment. The airspeeds, altitudes, and flight loads in a fighter airframe would provide more meaningful test data, and development of the high-performance engine could be accelerated. The Air Force liked the idea, so Weaver convinced the Navy to bail both Skyrays to the Air Force, which, in turn, bailed them to General Electric.

Weaver decided to use 587 as the flying test bed and cannibalize 586 for parts. The airframe of 587 was significantly modified for the new engine. The J79 was shorter and lighter than the Westinghouse J40 it replaced. New mounts were fabricated, and new inlet ducting was designed. General Electric technicians had difficulty forming the ducts so Weaver hired California Metal Shaping, a custom automotive body company, to hand form the inlet ducting. A new tail cone was installed to accommodate the J79 afterburner nozzle, and the original Skyray instrument panel was replaced with a panel rich in engine instrumentation. Finally, the white paint was removed from 587 and its aluminum skin was given a high polish.

GE test pilot, Roy Pryor, made the first J79 flight on 8 December 1955. "The airplane had fantastic acceleration and performance compared to anything I have ever flown before, it felt like I had a tiger by the tail," recounted Pryor after the flight.[18]

Bill Todd, another GE test pilot remembered the characteristics of the J79 powered Skyray, "The control system wasn't very sophisticated. It was probably the worst control system I have ever flown. It was oversensitive, and there weren't proper null points. It wasn't relative, from how far you moved the stick to where you got a reaction...it hit you all of a sudden."[19] His assessment was very similar to Larry Peyton's after the first flight of 586 in 1951. However, Bill Todd learned to deal with the sloppy controls and flew the J79-powered Skyray for two years.

CJ805-3 engine being installed in BuNo 587. This was the commercial variant of the J79 intended for airliner application. (General Electric via Virgil Weaver, Nick Williams Collection)

BuNo 587 in flight with the CJ805-3 engine in April 1958. The aircraft has received a liberal dose of DayGlo paint on its nose, spine, and fin. The CJ805-3 flights used 300-gallon drop tanks for extended range. (General Electric via Nick Williams Collection)

29 April 1959, late in the CJ805-3 test program test pilot Elliot See (left), flight test manager Virgil Weaver (center), and flight test engineer Frank Randolf (right) discuss the results of a flight. (General Electric via Virgil Weaver, Nick Williams Collection)

BuNo 587 at Edwards late in its flying career. A wooden mock-up of the CJ805-23 aft fan engine is mounted on the right pylon. This XF4D-1 would be retired after these tests, having flown for nine years with no mishaps. (General Electric via Virgil Weaver, Nick Williams Collection)

The first J79 produced 13,500 lbf in afterburner, while an updated version reached 14,700 pounds. The XF4D-1 was now flying with nearly three times the thrust of its first engine—the 5,000-pound Allison J35. For two years the Skyray airframe absorbed the power of the J79 during 297 test flights totaling 202 hours.

In early 1958 General Electric removed the J79 and replaced it with its first commercial engine, the CJ805-3. This was a non-afterburning engine based on the J79 and was destined for the Convair 880 jetliner. It flew in 587 from 12 April 1958 until May 1960, accumulating 1,492 hours in 723 flights. The CJ805 test program was quite different than the J79 tests remembers Bill Todd, "Most of the early J79 flights were made clean—without drops [tanks]. The flight would be limited to 1.0 to 1.2 hours. The fuel specifics of the J79 were better than the J40, so we had no problem flying on internal fuel. The CJ805 tests required that we fly a transport profile, holding constant power settings at a constant altitude for long periods."

Todd found this accelerated service test program very boring. The tests were flown with 300-gallon drop tanks giving the XF4D-1 more than 2½ hours of endurance. The small cockpit was not very comfortable for long flights—the ejection seat always induced numbness—and adding to the discomfort was the Skyray's marginal stability requiring the constant attention of the pilot. To reduce the control system sensitivity, Todd would often extend the control stick and cruise in manual control. The greatest deficiency of the aging XF4D-1 was its lack of pressurization. "At 35,000 feet," said Todd, "the cockpit altitude was 31,000." On one flight Todd suffered the bends and had to land early.[20]

Finally, General Electric used 587 in a series of tests carrying a wooden scale replica of an *aft fan* engine, the CJ805-23. The engine model was instrumented with pressure probes and was mounted on the right pylon in place of the 300-gallon drop tank. The model was carried by 587 more than 20 times, providing data in lieu of wind tunnel tests.

At the conclusion of the CJ805 test flights, 587 was returned to the Navy. This prototype Skyray had been employed for years in rigorous testing and remained fully airworthy until the end. It is now on display at Naval Air Weapons Station (NAWS) China Lake, California. The first prototype, 586, which became a parts source for 587, was destroyed at China Lake while being used for firefight training.

XF4D-1 Ship 587 is seen on static display on the ramp at the Edwards Air Force Base Open House in October, 1959. (Nick Williams Collection)

The first prototype, BuNo 586, in 1961 at China Lake where it was used for fire fighting training. (Nick Williams Collection)

Famed NACA test pilot A. Scott Crossfield tests secret David Clark pressure suit he would use while flying the Douglas D-558-II Skyrocket. Cockpit shown here in pressure chamber at El Segundo was salvaged from the first XF4D-1 at China Lake before it was scrapped. Crossfield became the first man to fly Mach 2 in the Skyrocket. (U.S. Navy via David Clark Co. archives)

XF-92A Flight Test

The XF-92A arrived at Muroc on 9 April 1948 to prepare for its initial flight. The objectives of the flight test program were to determine the aerodynamic characteristics of the delta wing, assess stability and control characteristics, establish structural design criteria, determine aircraft performance, and evaluate the hydraulic flight control system, fuel system, and trim system. Convair test pilot Ellis D. "Sam" Shannon commenced the program by conducting a series of taxi tests, which disclosed brake and directional control problems. High-speed ground runs, which reached velocities of 165 mph, showed that the airplane was very sensitive in both pitch and roll control.

A lift-off in ground effect confirmed the control issues, so Convair engineers attempted to dampen the sensitivity. At the same time the engine was replaced with a slightly more powerful J33-A-23 (4,250 lbf) and the original clear canopy was replaced with a braced canopy. The work was completed on 15 September 1948 and test work continued. Additional high-speed taxis and lift-offs showed that the control system was still very sensitive, but by 17 September Shannon was satisfied that the XF-92A was ready for full flight. On 18 September 1948 Sam Shannon flew the world's first powered delta-winged aircraft, the XF-92A, almost 2½ years before Larry Peyton's first flight in the XF4D-1.

Shannon reported that the flight was uneventful, but he did experience a certain lag and sensitivity in the control system. After a thorough inspection that lasted 10 days, the XP-92A was released for its second flight on 29 September, and it flew every few days for the next several months. After the twentieth flight the full span elevons were replaced with shorter span surfaces in an attempt to dampen the overly sensitive control system. Convair test pilot William "Bill" Martin reported severe vibration above 399 mph. This condition was reproduced on several successive flights, so the full-span elevons were reinstalled. By 6 August 1949 Convair had completed all of its Phase I tests, which consisted of 47 flights totaling 20.5 hours.

On 13 October the Air Force began Phase II performance testing when Maj. Charles E. "Chuck" Yeager flew the XF-92A for the first time. He was joined by Maj. Frank K. "Pete" Everest in evaluating the control and stability of the delta wing. These tests were concluded on 28 December 1949 after 25 flights totaling 17 hours. The aircraft was returned to San Diego for maintenance and modification.

The Air Force, encouraged by the early flights, issued a contract to Convair for the installation of an afterburning Allison J33-A-29 engine. This required significant modification to the airframe to accommodate the longer engine. Additional fuel cells were installed, the fuselage was lengthened, and the natural aluminum airframe received a coat of white paint to aid optical tracking. The new engine provided 5,200 lbf in military power, and 7,500 pounds in afterburner, an increase of 3,250 pounds over the prior engine. Convair estimated that the new engine would increase maximum speed at sea level from 552 to 718 mph. At 30,000 feet the speed would increase from 589 to 669 mph. The rate-of-climb at sea level would increase from 5,580 to 10,650 fpm. The takeoff distance would decrease from 5,400 to 3,900 feet, and the time-to-climb to 30,000 feet would decrease from 12.4 to 3.4 minutes.

The rework and engine change took 14 months. On 21 July 1950 Chuck Yeager made the first flight with the new engine. Yeager and Everest made 21 flights totaling 11 hours and 21 minutes, but the performance with the afterburning engine was disappointing. The XF-92A was only able to achieve supersonic flight once, on 2 February 1953, when Pete Everest flew it in afterburner in a prolonged dive. The XF4D-1 Skyray had flown supersonically six months earlier in the spring of 1952 without the benefit of an afterburner.

The NACA High-Speed Flight Station at Edwards took possession of the XF-92A after Everest's supersonic flight. It was flown by NACA test pilot A. Scott Crossfield 25 times between 9 April and 14 October 1953 for a total of 13 hours and 13 minutes. NACA installed yet another Allison engine, the J33-A-16 that produced 8,600 lbf in afterburner. Now the XF-92A was flying with more than twice the power of its original engine. Crossfield discovered a severe pitch-up tendency in high-speed turns. Wing fences were installed, but they only cured the problem partially. On 14 October 1953 Crossfield suffered a collapsed nose gear on landing at Edwards. The damage was minor, but NACA decided to permanently ground the XF-92A since the follow-on YF-102 was scheduled to fly in 10 days. The XF-92A made 118 flights for a total of 62 hours; it is currently on display at the Air Force Museum in Dayton, Ohio.[21]

The XF-92A during its initial test flights. This was the world's first delta-winged aircraft to fly, making its maiden flight on 18 September 1948, nearly 2½ years before the XF4D-1 Skyray. (Convair via the San Diego Aerospace Museum Archives)

The XF-92A in July 1950 after the installation of an afterburning Allison J33-A-29 engine. The engine required a lengthening of the fuselage. The aircraft was painted all white to improve visibility. (Convair via the San Diego Aerospace Museum Archives)

The XF-92A suffered a nose gear failure on 14 October 14 1953 while flying for the NACA High-Speed Flight Station. The damage was slight, but the aircraft was never flown again. (Convair via the San Diego Aerospace Museum Archives)

Chapter Three

The Engine Story

The business end of the F4D-1's Pratt & Whitney J57-P-8A is visible here. This engine produced 16,000 pounds of thrust in afterburner, nearly 60 percent more than the Westinghouse J40-WE-8 that was initially specified for the Skyray program. (Nick Williams Collection)

Jet engine development had a profound effect on the evolution of the Skyray. To fully understand the history of the airplane, it is helpful to examine the corporate histories of two engine manufacturers that were involved in the program, Westinghouse and Pratt & Whitney, and their relationship with the Navy Bureau of Aeronautics (BuAer).

Westinghouse

Westinghouse was present at the birth of the American jet engine industry. In April 1941 three companies, Westinghouse Electric, General Electric, and Allis-Chalmers were invited by the government to investigate the new concept of gas turbine propulsion for aircraft. These companies were selected because of their experience with steam and gas turbine technology. Traditional aviation engine manufacturers were excluded since it was feared that they would resist exploring a technology that could compete with their existing piston-engine product lines.

General Electric was chosen to conduct research for the Army, while Westinghouse was selected to work for the Navy. Allis-Chalmers was given a Navy contract for a ducted-fan turbojet, but failed to produce promising results and the contract was canceled. Westinghouse and General Electric achieved success, but the path of the two companies couldn't have been more divergent. The Army asked General Electric to manufacture an existing design, the Whittle turbojet, a British engine that had already flown in the Gloster E28/39. Westinghouse, however, was given a free hand by the Navy to design and build an original engine.

Westinghouse elected to design an axial flow engine as opposed to the Whittle/General Electric centrifugal flow. This allowed Westinghouse to develop a narrow-diameter 19-inch, 850-pound engine that produced 1,200 lbf. The General Electric engine was 41.5 inches in diameter and produced a similar amount of thrust. Westinghouse built its engine in just 16 months and designated it the Model 19A. It was nicknamed the Yankee to emphasize its all-American heritage. The Yankee was flown on a Vought FG-1 Corsair (mounted under the fuselage) on 21 January 1944, impressing the Navy with the rapid progress. Westinghouse had developed an axial flow engine from scratch that was considered superior to the Whittle design in many ways. Ultimately the Yankee matured to become the J30 and was selected to power the McDonnell FD-1 Phantom, the Navy's first jet-powered carrier fighter.

The General Electric 1-A turbojet engine. A direct copy of the Whittle W.1X from England. It ran on 18 April 1942, making it the first jet engine to operate in the United States. The engine was 72 inches long, 41.5 inches in diameter, and produced 1,250 pounds of static thrust. Two 1-A engines powered the Bell XP-59A on its initial flights. (Aircraft Engine Historical Society)

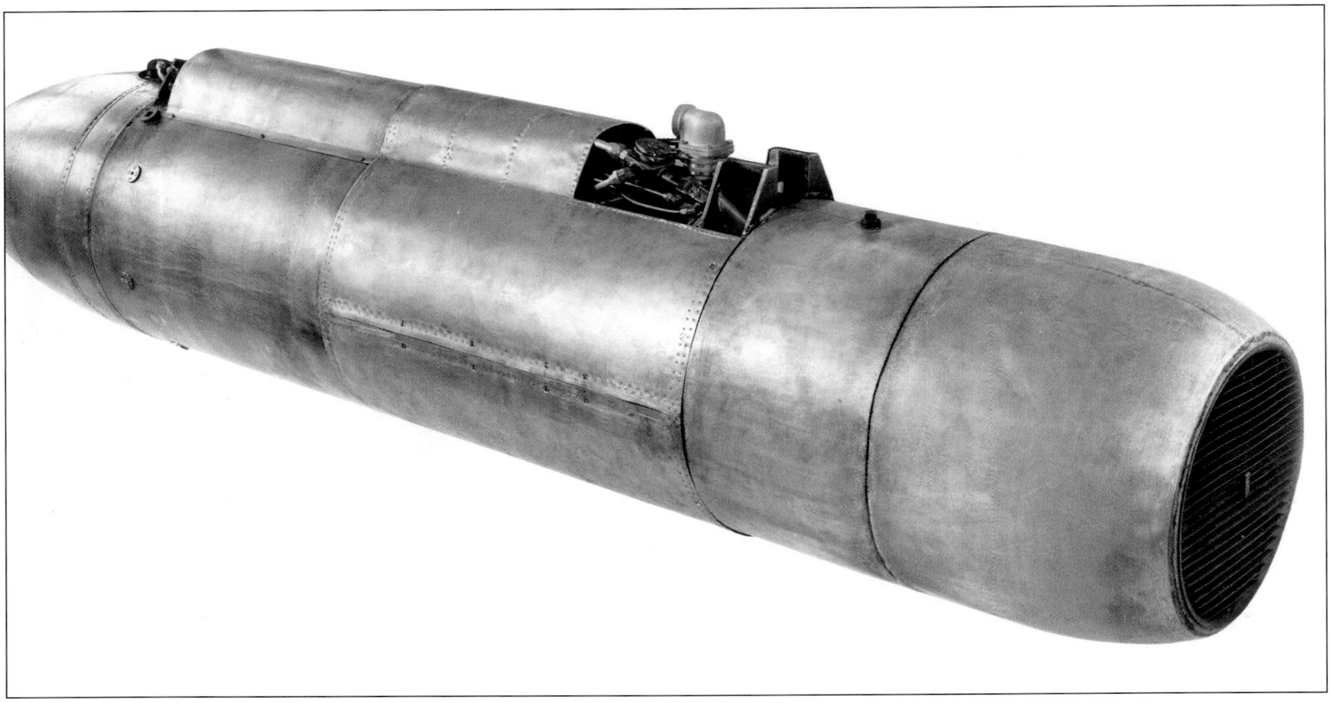

The Westinghouse 19A Yankee was the first American-designed gas turbine engine. It used an axial flow compressor and was only 19 inches in diameter. (Westinghouse via the Hagley Museum Collection)

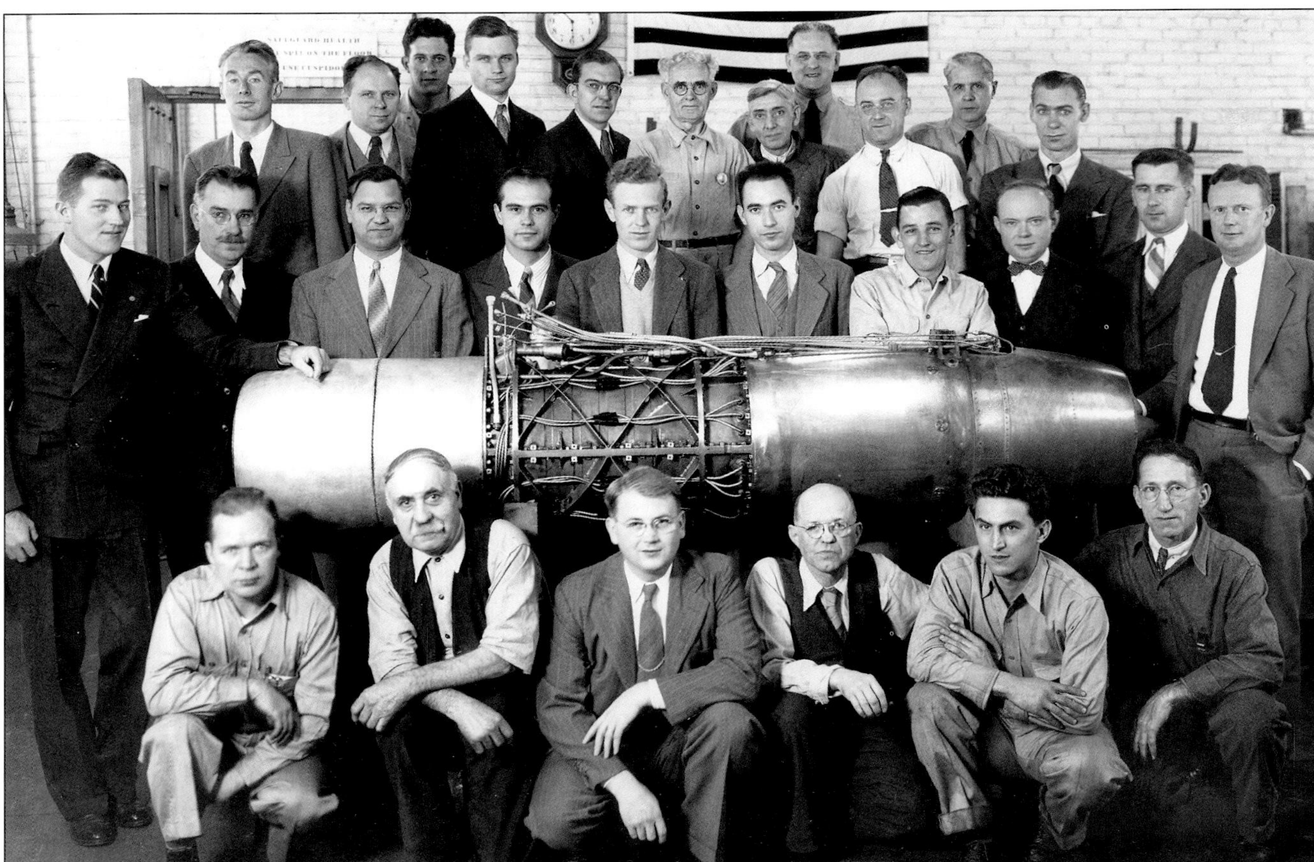

This photograph was taken on 7 October 1943 showing the Westinghouse staff responsible for producing the 19A engine, the first gas turbine engine designed in the United States. (Westinghouse via the Hagley Museum Collection)

But problems began to surface when Westinghouse was faced with mass production of the J30. Historian Paul D. Lagasse stated that Westinghouse was ill prepared for large-volume production. The engineering culture of the Turbine Division was based largely on steam turbines, which are manufactured one at a time to unique specifications. Furthermore, the Division's manufacturing facility was a cramped space in South Philadelphia that did not allow for efficient production lines. BuAer was painfully aware of Westinghouse's inability to produce large numbers of the J30 so they turned to Pratt & Whitney, a company noted for its outstanding manufacturing prowess. Therefore, as Lagasse notes, the first quantity production order for Westinghouse aircraft gas turbine engines was not to be filled at Westinghouse. BuAer issued Pratt & Whitney the prime contract for 500 engines and Westinghouse was issued a sub contract for only 50 engines. Pratt & Whitney's capabilities proved

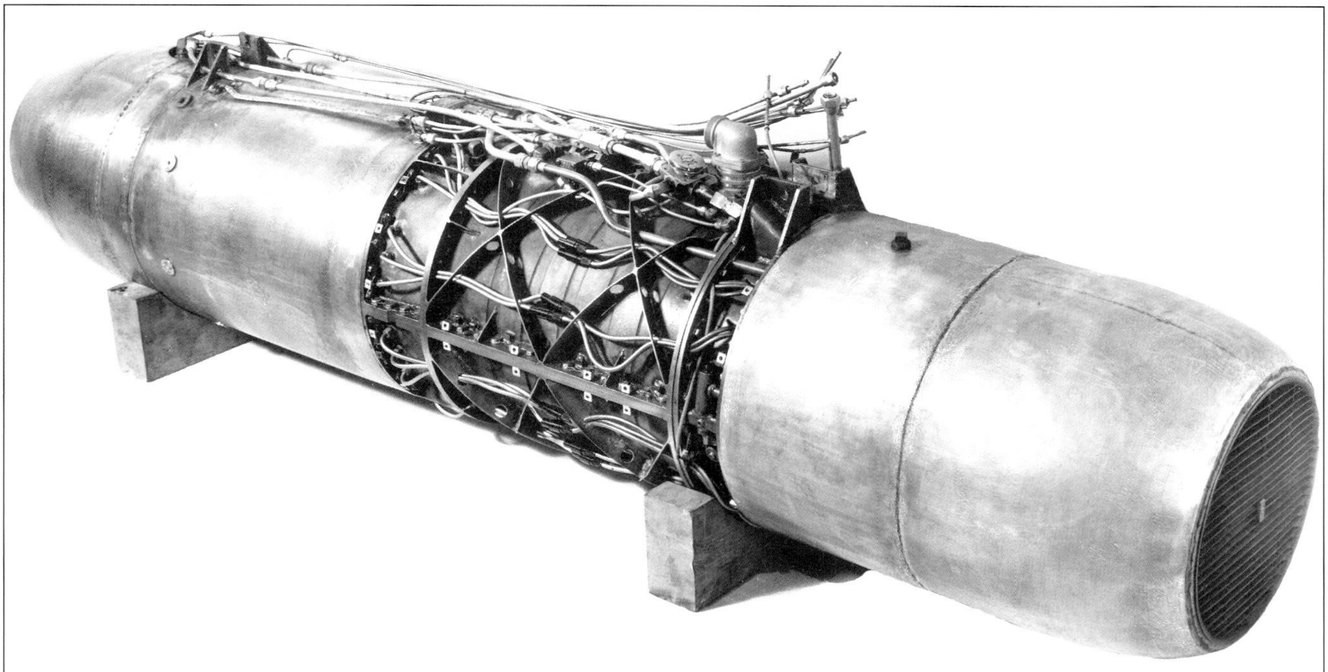

The Westinghouse 19A Yankee showing the interior engine casing, which was covered with a smooth exterior skin. This engine was flown under the fuselage of a Vought FG-1 Corsair. (Westinghouse via the Hagley Museum Collection)

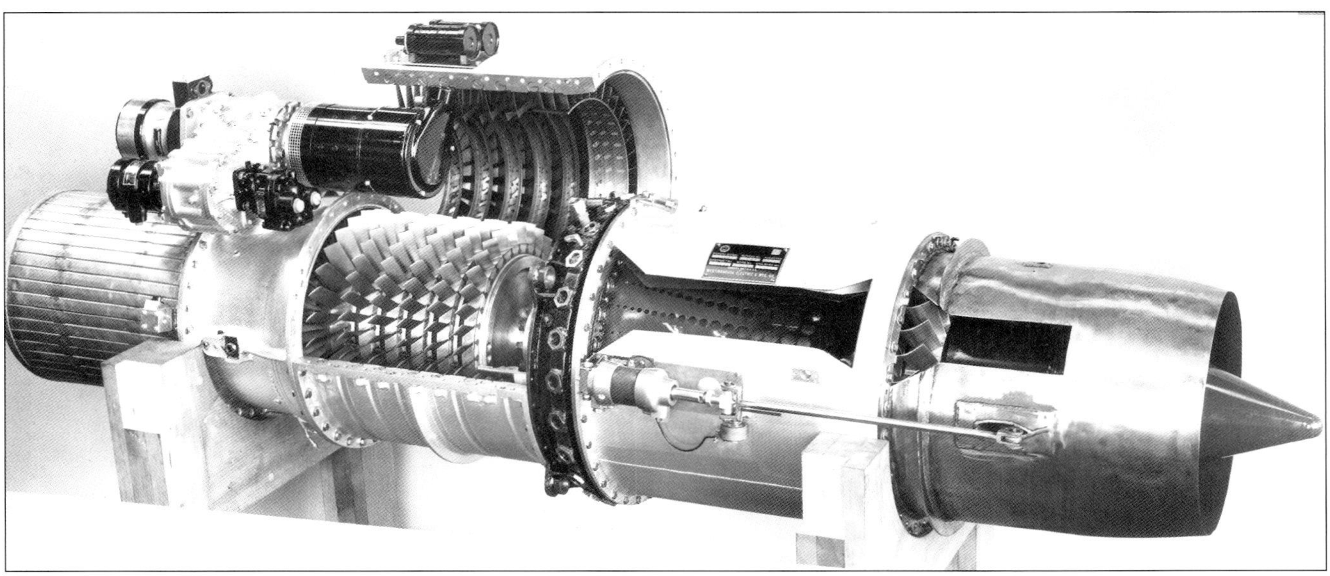

A cutaway display of the 19B engine showing the multi-stage compressor section. This engine would evolve into the J30, which powered the McDonnell FD-1 Phantom, the first jet aircraft to operate off a carrier deck. (Westinghouse via the Hagley Museum Collection)

A public display in June 1945 of the three gas turbines developed by Westinghouse. The small engine in the middle was the X9.5B, an experimental half-sized version of the 19A. (Westinghouse via the Hagley Museum collection)

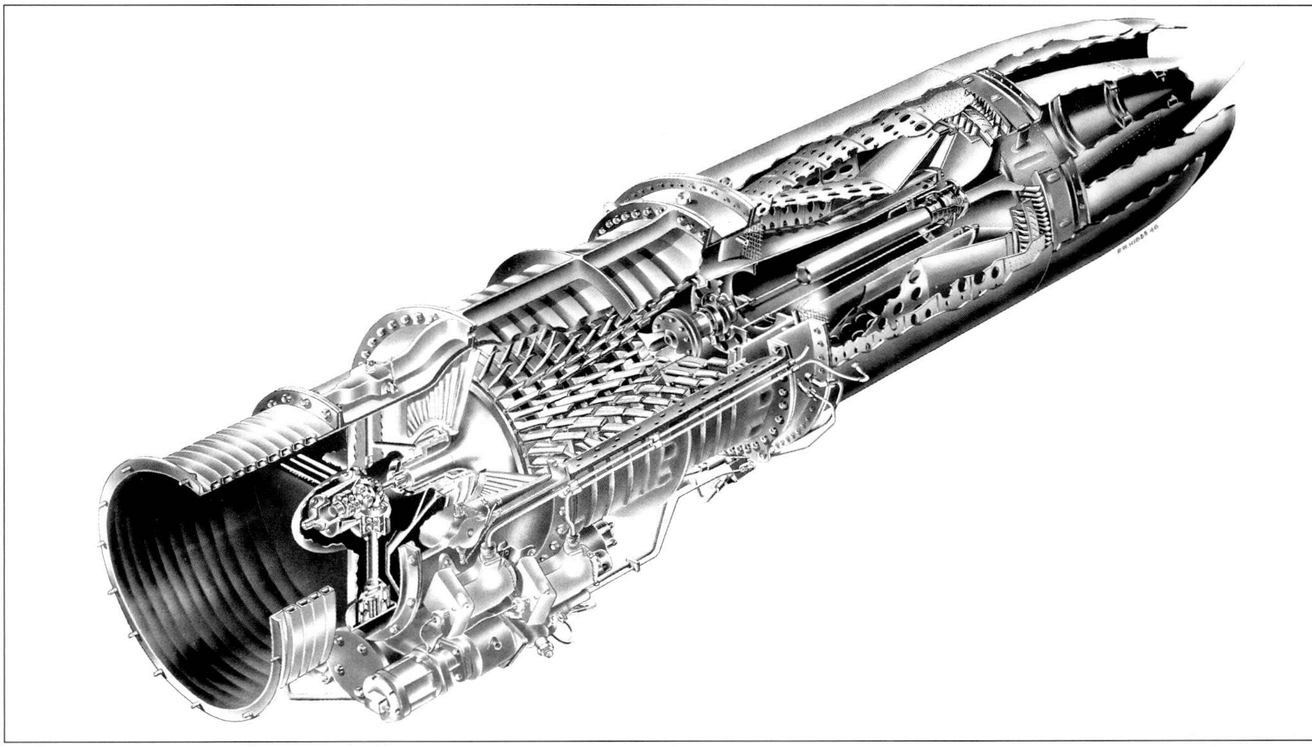

A cutaway drawing of the 19XB-2B, which received the designation J30. It was the first Westinghouse engine in serial production. However, Pratt & Whitney produced most of the units under license. (Westinghouse via the Hagley Museum Collection)

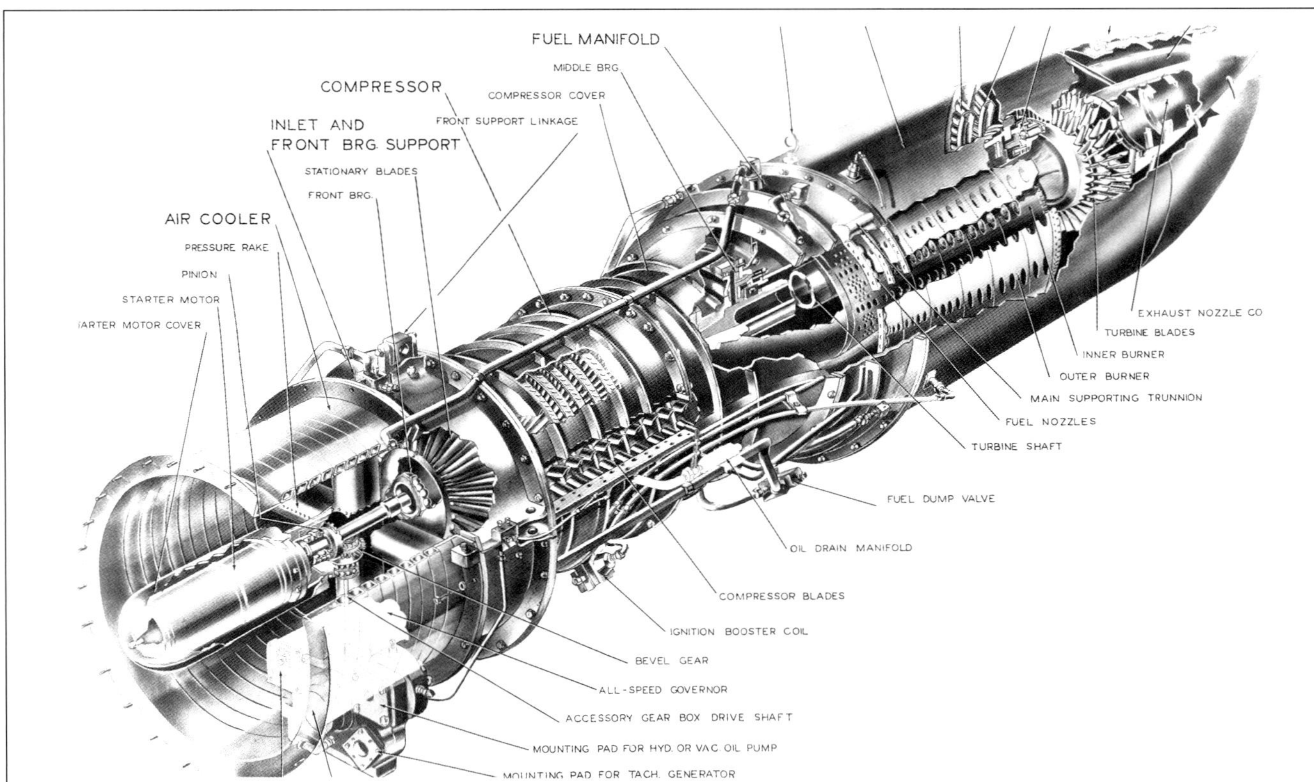

A cutaway drawing of the 24C, a scaled–up version of the 19B engine. This engine was designated the J34, and became the most successful Westinghouse turbojet. By the early 1950s Westinghouse was regarded as the Navy's leading jet engine vendor. (Westinghouse via the Hagley Museum Collection)

A display model of the Westinghouse 24C axial flow turbojet engine. This design was an enlarged version of the original Westinghouse turbojet, the 19A. It first ran in April 1945 producing 3,000 lbf, and it was flown in the Chance Vought XF6U-1 Pirate. Ultimately the 24C became the J34, one of the most successful early turbojet engines. (Westinghouse via the Hagley Museum Collection)

A side view of the Westinghouse J34, which powered many early Navy jets. The J34 produced 3,000 lbf and weighed 1,255 pounds. It could be fitted with an afterburner to produce 4,000 lbf. (Westinghouse via the Hagley Museum Collection)

A cutaway display of the Westinghouse J34 turbojet photographed in August 1953. The J34 became the most successful engine produced by Westinghouse, establishing Westinghouse as the Navy's leading jet engine vendor. (Westinghouse via the Hagley Museum Collection)

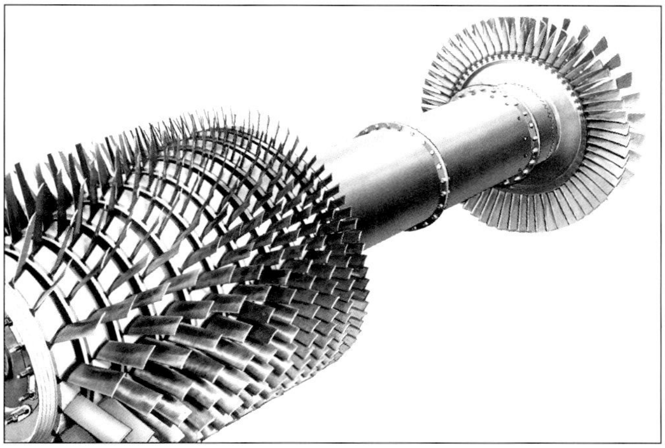

The J40 required a very dense compressor to achieve the proper pressure ratios necessary for high thrust. This became the most technologically demanding component of the engine. (Westinghouse via the Hagley Museum Collection)

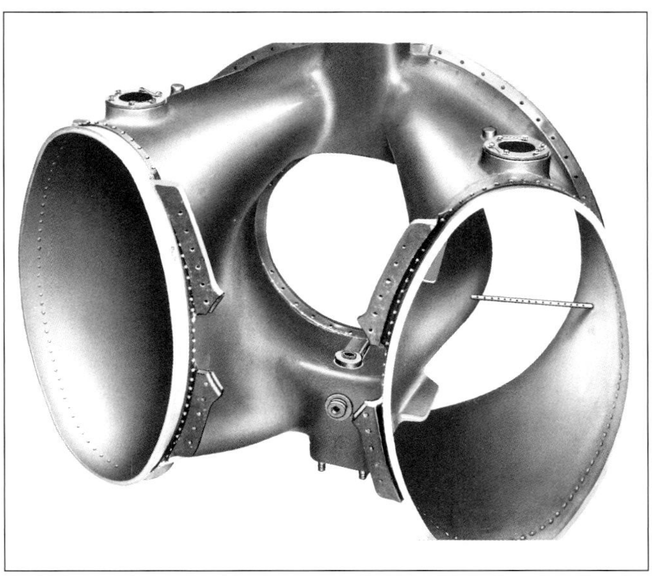

The J40 featured a bifurcated inlet for the Skyray installation. The engine accessories and gear box was mounted on the engine centerline between the air inlets. (Westinghouse via the Hagley Museum Collection)

well suited to the manufacture of jet engines, and they gained valuable experience building the Westinghouse design.[1]

In 1947 Westinghouse scaled up the J30 design, which evolved further to become the J34, the powerplant for McDonnell's F2H-1 Banshee. By this time the Navy held Westinghouse turbojet designs in high esteem, and they expected that Westinghouse could overcome the production difficulties. The J34 became the engine of choice for many of the aircraft in the Fleet, and the Navy helped Westinghouse production relocate to a sprawling manufacturing plant in Kansas City.

Initially J34 production went well. In January 1950 the Kansas City production line delivered its first J34 to the Navy, and by September 1951 the 3,000th engine was completed. Westinghouse was producing J34s at a rate of 150 per month, with plans to double the rate. BuAer was convinced that Westinghouse had come of age, and it was regarded as the Navy's primary jet-engine vendor. By the early 1950s Westinghouse had two new engines in development, the J40 and the J46. Between them it was expected that Westinghouse engines would power almost all Navy fighter and attack aircraft. Four new Naval aircraft were designed around the J40: the F4D Skyray, the A3D Skywarrior, the F3H Demon, and the Grumman F10F Jaguar.

To the Navy, powerplant development seemed to be the least of its problems. Westinghouse had successfully doubled the thrust of its J30 with the J34, and the J40 was expected to be a relatively low risk expansion of the same design. Westinghouse committed to produce the engine with more thrust in less time and at lower cost than any of its competitors. Given Westinghouse's track record, it was an easy decision for the Navy to specify J40 power for its new airplane programs.

Soon the Navy would regret its decision. Westinghouse retained its development and testing operation in Philadelphia, nearly half a continent away from its production facility. Early J40 development seemed to go well, the engine was first run in late October 1948, and BuAer received a report from LCDR Neil E. Harkleroad stating, "The engine has been operating successfully to date."[2] However, the J40 proved to be a bitter disappointment. Not only did Westinghouse underestimate the difficulty in growing the basic 4,200-lbf J34 to 7,400 pounds in the up-scaled J40, but BuAer asked for more thrust for better fighter performance. Westinghouse agreed to a more powerful engine that would require a 14-stage compressor with an enhanced turbine. This proved to be too great a leap for Westinghouse's resources. The J40 program became a debacle of major proportions. Severe stress was placed on Westinghouse engineers, technicians, managers, and suppliers. Development in the South Philadelphia facility was constrained by space and equipment limitations. Even the availability of engine test stands became a problem. The facility faced the impossible task of trying to fix flight test issues on the new J40, while meeting production demands for four Navy and two Air Force airplanes scheduled to receive the engine.

Numerous problems emerged during J40 flight-testing. Rotating stall was the result of the engine's densely packed compressor section. A blade stall on one compressor blade would affect the aerodynamics of the neighboring blade. This would set up a vibration around the circumference of that stage that resulted in high blade loads and failure, followed by engine failure.[3]

Grumman test pilot Corky Meyer recalled that the J40 used in the XF10F flight test program required 21 seconds to accelerate from idle to full thrust. Other engines of that period required only 5 to 6 seconds.

The interior of the XF4D-1 engine bay looking forward. These air ducts mated to the J40 compressor inlets. (Westinghouse via the Hagley Museum Collection)

A mock-up of the J40 showing the bifurcated compressor inlets and the accessory components mounted between the inlets. (Westinghouse via the Hagley Museum Collection)

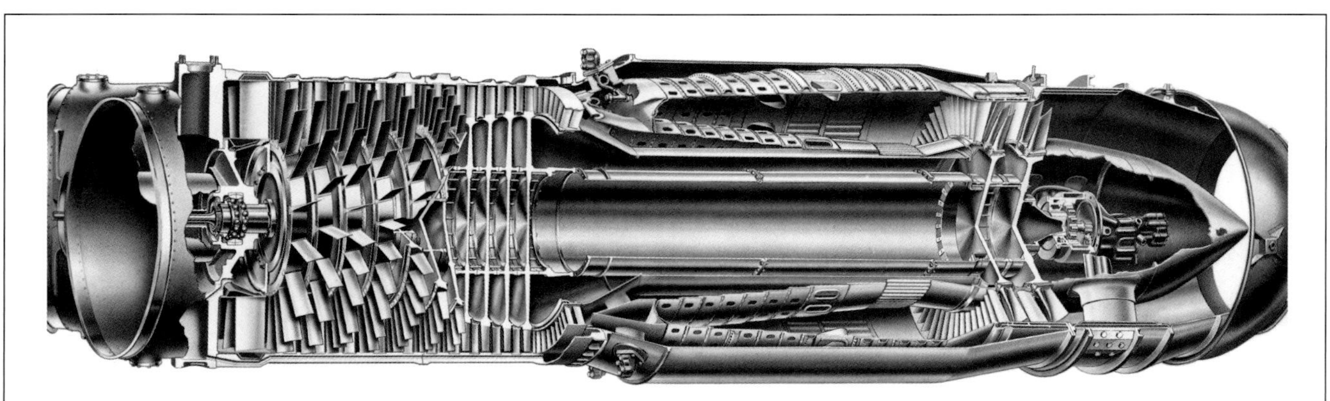

A cutaway drawing of the non-afterburning J40-WE-6 core engine. (Westinghouse via the Hagley Museum Collection)

A J40 being assembled at the Westinghouse plant in Kansas City. (Westinghouse via the Hagley Museum Collection)

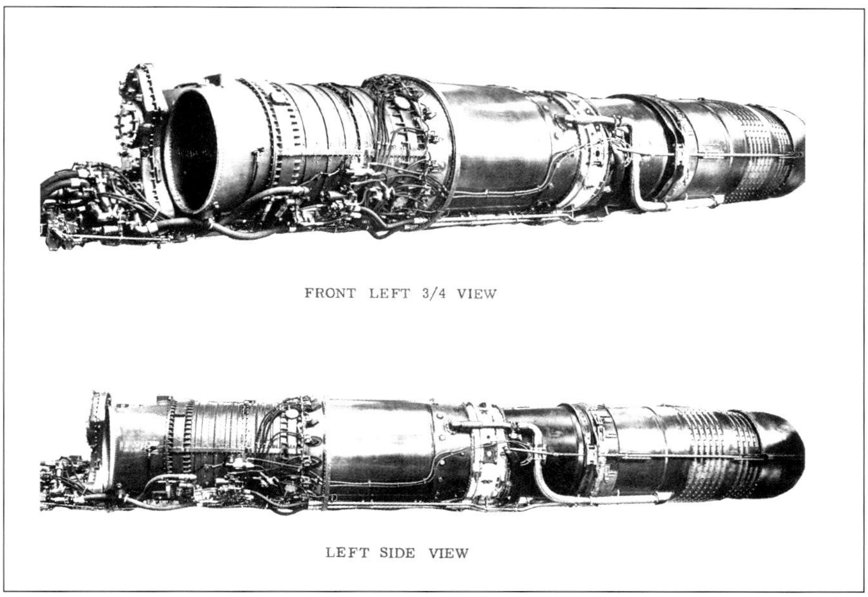

A front and side view of the afterburning J40-WE-8 intended for the Skyray. (Westinghouse via the Hagley Museum Collection)

Therefore, the wave-off performance of the J40 was abysmal. The following is Meyer's account: "The Westinghouse J40 was an interesting enough powerplant, being very heavy and operating at appreciably lower temperatures than those of its contemporaries. It had big bearings and castings and looked like the turbine for an electric locomotive or a hydroelectric station. If the J40 can be taken as a guideline, Westinghouse was hardly aviation oriented...

"After starting up the J40, one knew when the engine warmed up because the airplane stopped vibrating, smoothing out after a couple of minutes running. Every once in a while, my chase pilot would announce that a big puff of black smoke had been emitted from my tail. On the first occasions that this happened, I anxiously inquired if he could see any fire, but after getting a negative response several times, I realized that this was simply the way the J40 worked. From time to time it would coke up and expel a big blast of black smoke—all very unsettling but apparently not dangerous.

"One thing about the J40 that was more disturbing, however, was the extremely protracted acceleration time. The engine would accelerate to 100-percent RPM, and with further throttle movement, the clamshell would then close and start giving noticeable thrust, but before this happened 100-percent RPM corresponded to only 30-percent military [i.e., maximum non-afterburning] thrust. The engine had a 21-second

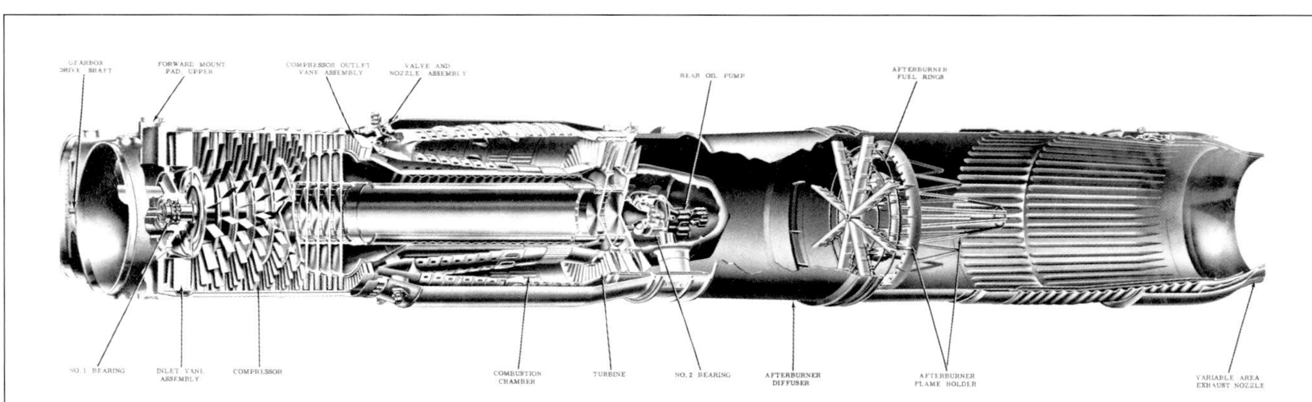

A cutaway drawing of the J40-WE-8 with afterburner. (Westinghouse via the Hagley Museum Collection)

acceleration time. If it was cut back to idling it took 21 seconds to get the engine up to speed once more and have the clamshell closed, and that can be an awfully long time under some circumstances."[4]

Summarizing the Westinghouse disaster, historian Anthony L. Kay said, "A prototype of the J40, an XJ40, was run on 28 October 1948 and an early J40 passed a 150-hour Mission Qualification Training (MQT) in January 1951. However, all was far from well with this turbojet and every type of problem involving combustion, aerodynamics, and mechanics manifested itself. Westinghouse was very unlucky to have so much wrong with an engine and the result was considerable unreliability and a failure to reach the design thrust of 3,405 kp (7,500 pounds) or 4,767 kp (10,500 pounds) with afterburner.

"...In January 1953 the Navy announced millions of extra dollars to increase the pace of development and get the engine right, but by March 1953 it had changed its mind and began seeking alternative engines from Pratt & Whitney and Allison for its aircraft. The J40 contract was canceled in September 1953, only 107 engines having been delivered."[5]

The J40 debacle didn't end with the cancellation of the contract. In September 1955 the congressman from St. Louis, Missouri's District, Representative Frank Karsten, took a tour of the McDonnell plant in St. Louis. There he saw 50 Demon airframes on the company ramp awaiting engines. He was told that the Demons were designed for Westinghouse engines that could not provide enough power to fly the airplanes safely. Karsten's visit resulted in a full congressional investigation into the procurement practices of the Navy.

Congressional investigators visited McDonnell, Westinghouse, and the Bureau of Aeronautics to interview senior staff at each. Representative Chet Holifield, chairman of the House Military Operations Subcommittee, held public hearings on 24 October to 27 October 1955 to sort out the responsibility for the failure of the Westinghouse-powered Demons. Westinghouse did not fare well at the hearings, and it became apparent that the Bureau of Aeronautics wasted more than $200 million on the J40 engine program. The Subcommittee concluded that the expenditures on the Demon/J40

A scene that became too familiar. This is the aftermath of a J40 component failure in a test cell. (Westinghouse via the Hagley Museum Collection)

The Westinghouse Aviation Gas Turbine engineering staff poses proudly with a J40-WE-8, the highest thrust engine of the time. (Westinghouse via the Hagley Museum Collection)

program were excessive, but that there was no evidence of dishonesty or improper influence. The Bureau of Aeronautics was criticized for not holding Westinghouse to a higher standard of performance and nine recommendations were issued to the Bureau for future compliance. Even though the Subcommittee imposed no civil or criminal penalties on Westinghouse, the damage to its reputation was fatal.[6]

Meanwhile General Electric and Pratt & Whitney were developing new jet engines that provided high thrust with excellent reliability. With its market share contracting, Westinghouse announced that it would concentrate on engines in the lower thrust range for trainers, helicopters, drones, and missiles.

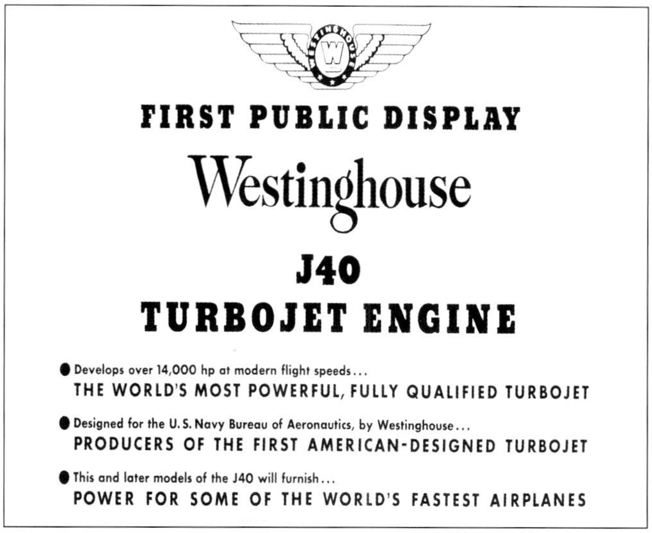

The J54, a mid-size engine with 6,000 lbf, was the centerpiece of Westinghouse's strategy. The engine was conservative in design featuring a 16-stage axial flow compressor with a two-stage turbine on a shaft supported by three bearings. To increase power and reduce fuel consumption, the compressor was designed to provide a compression ratio of 9:1—more than twice the 4.35:1 ratio of the successful J34.

Westinghouse was adventurous with the metallurgy of the engine planning extensive use of titanium, aluminum, and magnesium alloys for weight reduction. However, Westinghouse had little experience with these metals, so the prototype J54 was built with traditional steel. The engine completed its 50-hour endurance test within four months of its first test run, and validated the basic mechanical design. A concerted marketing program was launched targeting military and civilian operators, and some interest was generated. But the customers were eager to see the titanium engine perform, and unfortunately the first batch of forged titanium compressor blades were exhibiting defects. This delayed the test run of the titanium J54 by three months. This second J54 completed its 50-hour endurance test and was delivered to the Navy for evaluation in ground tests. However, the Navy declined to place a production order, which caused other customers to withdraw their interests.

Three years after launching the J54 program, Westinghouse had six engines in various stages of completion, but no sales. By 1956 Westinghouse had ceased to be a major manufacturer of jet engines. From 1956 to 1960 the Aviation Gas Turbine Division suffered constant

Westinghouse mounted a public relations campaign to repair the damage from its difficulties with the J40. (Westinghouse via the Hagley Museum collection)

Betty Furness, a noted actress, was a celebrity spokesperson for Westinghouse. She was used to advertise the virtues of the J40. (Westinghouse via the Hagley Museum Collection)

downsizing and loss of influence with BuAer. Occasional contracts were issued for work on the J34 engine. A new version of the engine was developed for the North American T2J-1 Buckeye trainer but only 217 examples of this single engine model were delivered before the Navy decided to purchase improved twin engine versions that were powered by General Electric or Pratt & Whitney engines. On 22 March 1960 Westinghouse announced that it would disband its Aviation Gas Turbine Division. All responsibility for servicing, engineering, and spare parts for Westinghouse engines still in service was given to Pratt & Whitney.

The 28 March 1960 issue of *Aviation Week* carried the news of the Westinghouse decision in a short paragraph: "Westinghouse Electric Corp. announced last week that it is withdrawing from the jet engine business and plans to return to the Navy the Kansas City, Missouri, facilities that house the firm's Aviation Gas Turbine Division. Westinghouse shares the Navy-owned Kansas City facilities with Bendix Corporation, which occupies approximately one-third of the floor space, and with a regional office of the Internal Revenue Service. Bendix is considering expanding into at least a portion of the plant space now occupied by Westinghouse, but the Navy said late last week that no final decision on this has yet been made."

Pratt & Whitney

Pratt & Whitney Aircraft Company (P&W) was founded in 1925 by Frederick B. Rentschler: a Princeton graduate, World War I veteran, and member of a powerful Hamilton, Ohio, industrial family. Rentschler's aviation interest grew out of his Army service where he was assigned to inspect and supervise aircraft engine production in the New York region. After the war Rentschler managed the newly formed Wright Aeronautical Company where he had a hand in developing the famous Wright Whirlwind. This was the engine that powered Charles Lindbergh's *Spirit of St. Louis* across the Atlantic.

In 1924 Rentschler left Wright and, with his family's connections, established his own aircraft engine company

The prototype XJ57 engine shown without an afterburner. The "wasp waist" contour of the engine case allowed the engine accessories and oil tank to be mounted close to the engine centerline, reducing the installation dimensions. (Aircraft Engine Historical Society)

jointly with the Pratt & Whitney Machine Tool Company of Hartford, Connecticut. That company, founded in 1860 by inventors Francis Pratt and Amos Whitney, joined forces with Rentschler and developed a reputation for producing reliable piston aero-engines such as the Wasp and Hornet. By the time the United States entered World War II, P&W was recognized as one of the premier aircraft engine builders in the world.[7]

The demands of wartime production left P&W with little time to consider the emerging technology of gas turbines. As an engineering exercise in 1941 the company began the preliminary layouts for a turboprop. A running prototype was built with $3.3 million of corporate funds, but corporate management regarded the turboprop as an expensive distraction and abandoned work on it in June 1945.[8]

The end of the war left little work for P&W. Overnight its orders shrank from $414 million to $3 million. It seemed that the industry was moving to gas turbine propulsion and P&W was well behind its competitors in that area. As we know, P&W had been awarded a contract to manufacture Westinghouse J30 engines when it became apparent that Westinghouse did not have the production expertise to mass produce its own design. This gave P&W its first real experience in the fabrication of gas turbine engines, but the company was still essentially a reciprocating aircraft engine company.

In 1947 P&W decided to enter the turbine engine business wholeheartedly. At that time the Navy needed an American source of mass production for the Rolls-Royce Nene turbojet to power the Grumman Panther. P&W had a reputation for efficient, high-quality manufacturing, but it was reluctant to build another company's design. The Navy induced P&W to produce the Nene by offering to pay all development costs including the license. With no risk, P&W found the offer too good to refuse and converted much of its factory space to jet-engine mass production. The license-built engine was designated the J42 and served as an important step in P&W's conversion to gas turbine manufacturing.[9]

The Pratt & Whitney J57 in production in November 1953. It required more than 20,000 tools to produce a J57 compared to the 5,250 tools used to produce the smaller J30. The J57 required 1,338,000 design man-hours compared to 730,000 design man-hours for the R-4360 Wasp Major. (Aircraft Engine Historical Society)

Pratt & Whitney executives examine an early J57 in November 1953. This was to become one of the most successful turbine engines in history. More than 21,000 units were produced. (Aircraft Engine Historical Society)

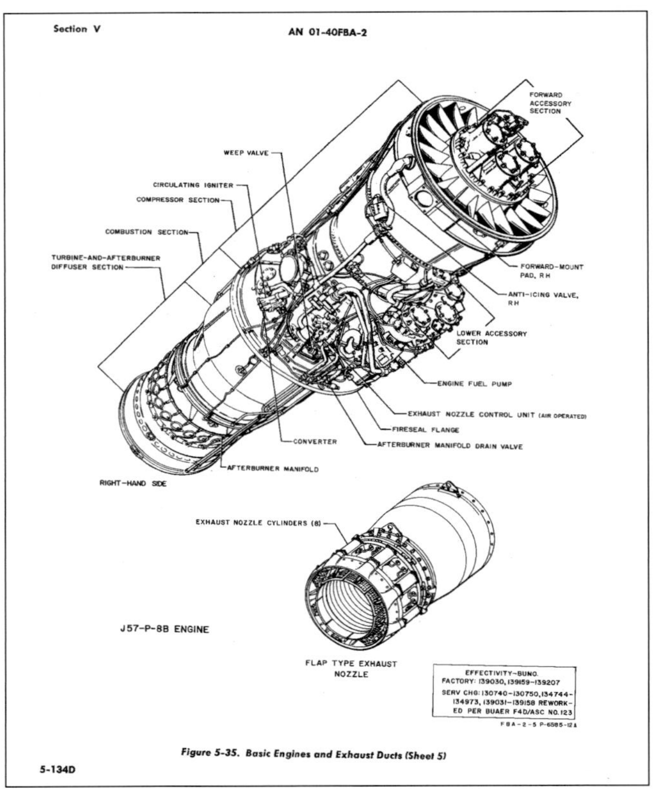

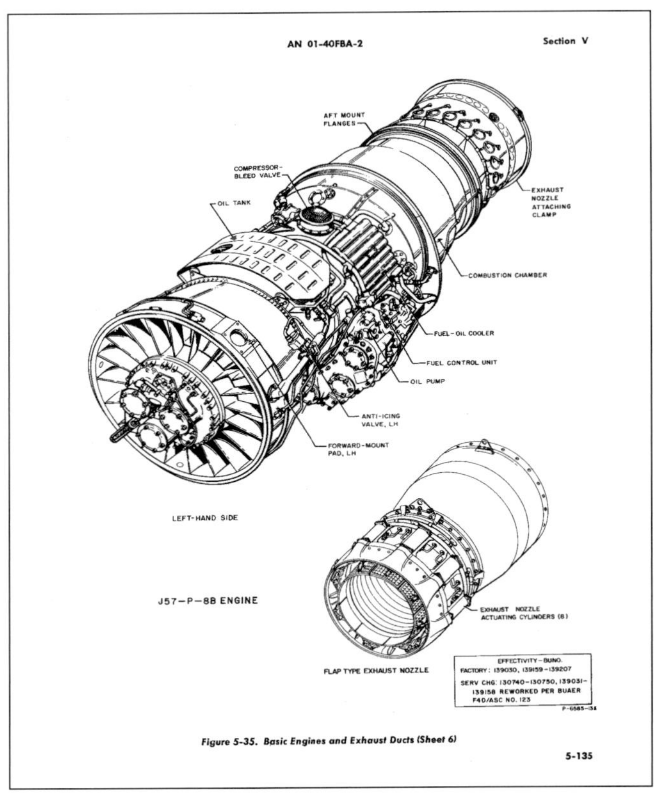

The Pratt & Whitney J57-P-8B engine shown in upper left side and lower right side views. This engine used the flap-type afterburner nozzle. (NAVWEPS 01-40-FBA-2)

Meanwhile, Rolls-Royce improved the 5,000-lbf Nene to become the 6,250-lbf Tay engine, which P&W manufactured in the United States as the J48. In 1950 P&W mated the J48 with an afterburner, increasing the thrust to 8,300 pounds, and successfully conducted the first aerial afterburning tests on the prototype Lockheed YF-94C. In 1954 the Navy flew a swept-wing version of the Grumman Panther—the Cougar—supersonically in a shallow dive with a P&W J48. This was one of the few occasions when a centrifugal flow turbojet flew beyond Mach 1.[10]

Pratt & Whitney's joint venture with Rolls-Royce resulted in about 5,250 engines produced for the United States Navy and Air Force. This license agreement enabled P&W to enter the jet engine business quickly and gain important experience with minimum risk. Pratt & Whitney had become an important manufacturer of gas turbine engines, but there was a growing awareness that the company was lagging behind the competition in the design and development of technologically advanced jet engines.

Pratt & Whitney's engineering manager, Leonard S. Hobbs, championed a twin-spooled axial flow design that promised remarkable performance and efficiency. Andrew Willgoos, who had been P&W's chief designer since the company was formed in 1925, conducted design work. The twin-spool concept involved a low-pressure and high-pressure compressor driven by matching turbines that were mounted on concentric shafts—one spinning inside the other. Pratt & Whitney's sponsor for this new engine was the Air Force, which needed an engine of high efficiency for its XB-52 long-range bomber. This new powerplant became the stalwart J57; however, early J57s were not trouble free.

Guy Townsend, a B-52 test pilot said, "I never had an eight-engine takeoff; always seven. You didn't know which seven, but they failed one at a time!"[11] The main problem was the vacuum tube electronic control system. This was eventually replaced with a hydro-mechanical control system, and the J57 became a model of reliability. Now the B-52 could be counted on to take off, cruise, and land with all eight engines running.

The Air Force acquired great confidence in the engine and specified the J57 for several of its new-generation fighters, the Century Series: F-100 Super Sabre, F-101 Voodoo, and F-102 Delta Dagger. The Air Force success with the J57 caused the Navy to look to P&W when the Westinghouse J40 was canceled. Both the Douglas A3D Skywarrior and the F4D-1 Skyray were modified to receive the J57. The Navy's next-generation fighter, the Vought F8U, was also designed around the J57 engine.

In its commercial form, the JT3 enjoyed enormous success powering many of the first-generation U.S. jetliners such as the Boeing 707 and Douglas DC-8. Historian Bill Gunston characterized it as the most important postwar jet engine. Ultimately more than 21,000 J57 or JT3 engines were produced.[12]

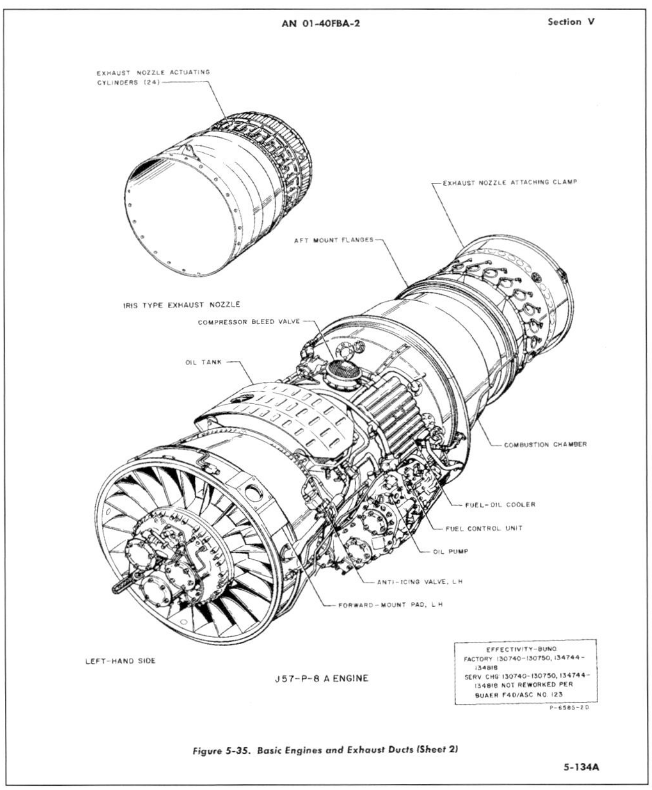

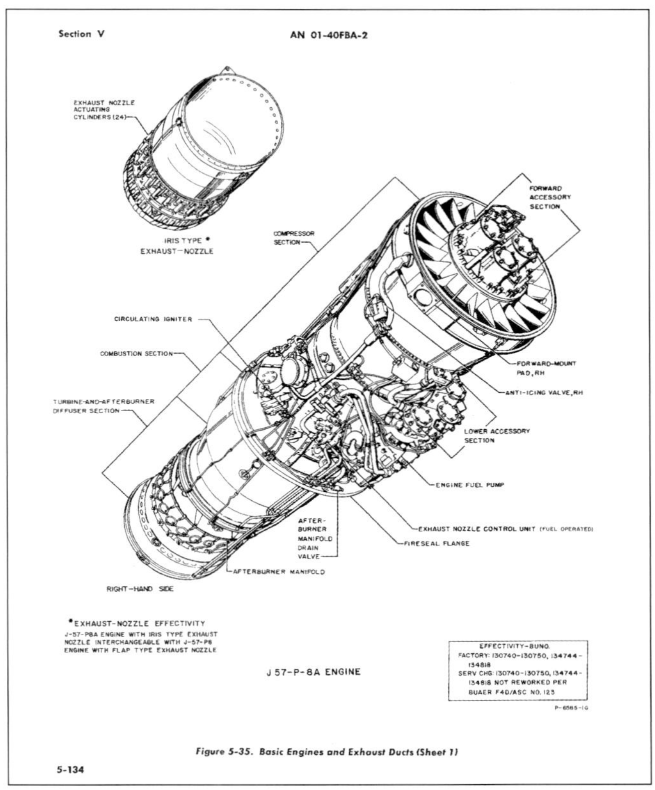

The Pratt & Whitney J57-P-8A engine shown in upper-left-side and lower-right-side views. This engine used the Iris-type afterburner nozzle. (NAVWEPS 01-40-FBA-2)

Chapter Four

Skyray Production

Compare this production configuration F4D Skyray to the first prototypes shown in Chapter One. New Douglas navigation pod is shown on the inboard pylon, and air intakes are now optimized by the use of a boundary layer splitter plate. (Nick Williams Collection)

Using new production methods for building a new airplane in a new factory posed obvious challenges to the engineering and manufacturing teams at Douglas Aircraft. As always, Douglas was up to the task.

The First 12

On 10 February 1951, little more than two weeks after the first XF4D-1 flight, the Navy issued Douglas a letter of intent for 12 production F4D-1s. By 25 October 1951 the start of drawing release for the first 12 production aircraft was underway. Recalling his surprise at the speed of the Navy's production order Heinemann said, "Even though the first two airplanes were experimental articles, soon after they had flown a few times and BuAer digested their performance data, we were asked how long it would take to put the Skyray into production. This was a bit of a surprise but was like asking a child if he or she would like a lollipop. 'Almost immediately,' was our answer."[1] Actually, it would take 3½ years for the first production Skyray to fly. BuAer's reliance on the Westinghouse J40 was the primary reason for the delay.

Heinemann had been leery of the Westinghouse J40 from the outset of the Skyray program. He insisted that the Skyray's engine bay be large enough to accommodate other engines in the event that the narrow-diameter J40 didn't work out. His concern proved to be well founded. The Westinghouse engine, when it was finally delivered, was extremely troublesome. Heinemann had suffered similar problems on the A3D Skywarrior project, which also relied on the J40. In a book he co-authored with Rene Francillon on the A3D Skywarrior, Heinemann wrote, "As early as January 1952 the Navy had become concerned with the difficulties experienced by Westinghouse in developing the J40 engine and considered the use of J57s to avoid delaying the urgently needed A3D. Two months later when it became clear that 7,500 pound (3,402 kg) thrust J40-WE-12s, as specified for the production A3D-1s, would not be available prior to the last quarter of 1954, the new P&W two-spool turbojet being developed for the Air Force became an attractive alternative and was given more serious consideration."[2]

On 2 April 1952 the Navy placed an initial order with P&W for a batch of J57-P-1 engines, and modified the Douglas contract to specify that engine for all production Skywarriors. The authors noted, "For Heinemann and his team, this decision represented a final vindication of their long advocated recommendation for the use of Pratt & Whitney J57s to power their heavy attack bomber."[3] By 15 March 1953 the Navy made a similar decision with regard to Skyray production. The afterburning variant, J57-P-2, was specified for the production Skyrays.

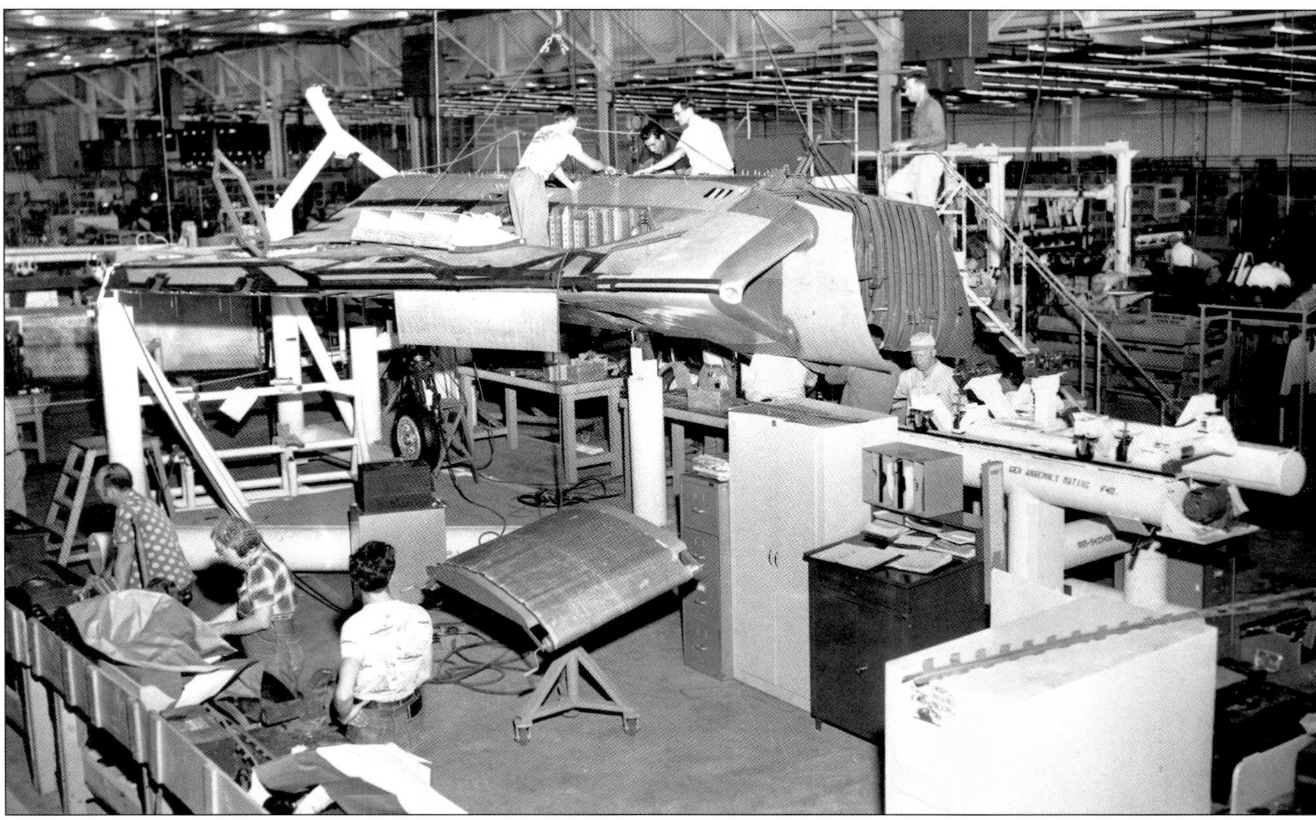

This series of photographs shows the first production Skyray, 130740, under construction in the fall of 1953. The first block of 12 aircraft were hand built. The production assembly line was established later. (Nick Williams Collection)

In April 1953 the Douglas design team commenced work on the detail changes needed for installation of the new powerplant. Nearly 80 percent of the airplane was affected by the redesign; several airframes, already under construction pursuant to the 1951 letter of intent, required significant retrofitting. Compounding the difficulties, Douglas was forced to establish a new assembly line to accommodate the production of Skyrays at a converted aluminum plant in Torrance, California. Backlogs from the Korean War had stretched the El Segundo plant to its limit. Nevertheless, the first production Skyray, BuNo 130740, was completed within 13 months of the engine change approval.

The aircraft that emerged from the production line were significantly different from the prototype XF4D-1s. Aside from the new P&W engine, the production Skyrays sprouted numerous scoops, access doors, louvers, landing gear door changes, cockpit changes, and tail cone changes. One of the most significant changes was the split rudder at mid-span. The upper half acted as a yaw damper while the lower half acted as a traditional rudder.

Bob Rahn made the first flight of the first production article, 130740, on 5 June 1954 from the Los Angeles International Airport. Rahn was delighted with the P&W engine that provided 14,500 lbf in afterburner, approximately 4,000 pounds more than the afterburning Westinghouse J40-WE-8. A new flight control configuration with a different force-feel system was employed, which was a dramatic improvement over the troublesome controls of the prototypes.

The new engine was a great enhancement to the Skyray and Rahn was able to achieve supersonic flight in level flight easily, but the inlet ducts were inadequate for the large engine. A series of duct revisions were investigated from October 1954 until August 1955 when a compromise design was agreed upon. A high-altitude compressor stall problem was never completely solved but it was minimized. As Bob Rahn recalled, "The J57 was a fine, reliable engine and that additional thrust over the J40 engine enabled the Skyray to be a true interceptor. However, as with anything else, there is bad with the good. Due to its delta-wing design and resultant low wing loading, the airplane was capable of pulling almost twice as many Gs at high altitude than present contemporary fighters. A compressor stall of the engine, however, almost always ensued, which negated the tactical advantage. We evaluated almost as many different engine duct inlets for the subsonic, high-altitude, high-alpha regime as we did tail cones for the high-Mach, transonic

Various views of the first production article, 130740. While this aircraft employed many features that would become standard on later production aircraft, it still had items that were carried over from the XF4D-1s such as the engine inlets. This Skyray was stricken after incurring significant structural damage from a 9.1-g pitch-up. (Nick Williams Collection)

speed range. We finally settled on the engine duct inlet design seen on production Skyrays."[4]

All of the first 12 production aircraft (BuNos 130740-130751) flew with the Douglas Testing Division except 130746, which went directly to the NATC for carrier evaluation, and 130751, which became a static test article. These Testing Division Skyrays were Navy owned, but placed on bailment to Douglas for experimental testing.

The first four aircraft were painted gloss white, while the others were painted glossy Sea Blue. Over time the blue aircraft received standard Gull Gray and white color schemes. BuNos 740, 742, and 744 were used as aerodynamic demonstration aircraft. BuNo 741 was used initially

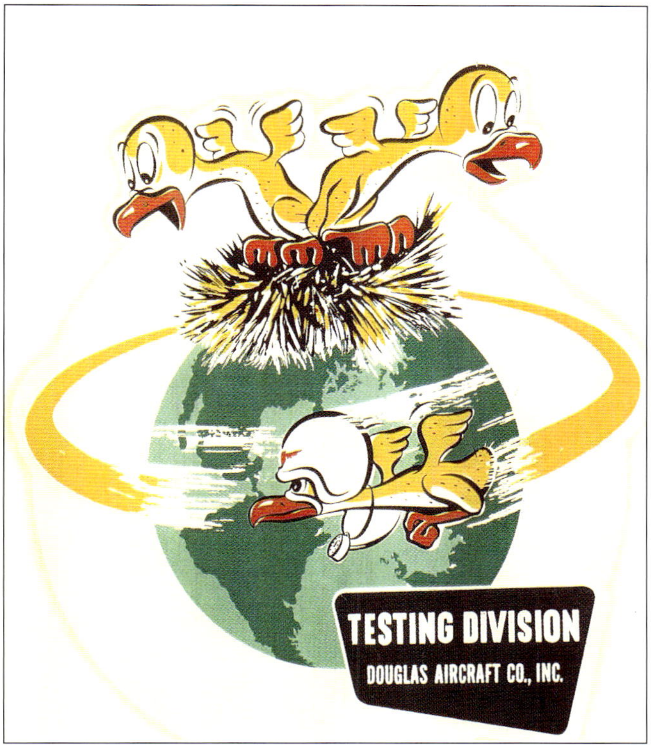

The Douglas Testing Division logo. This logo appeared on the fin of most Testing Division aircraft. It also appeared on pilot helmets and mechanic toolboxes. (Author's Collection)

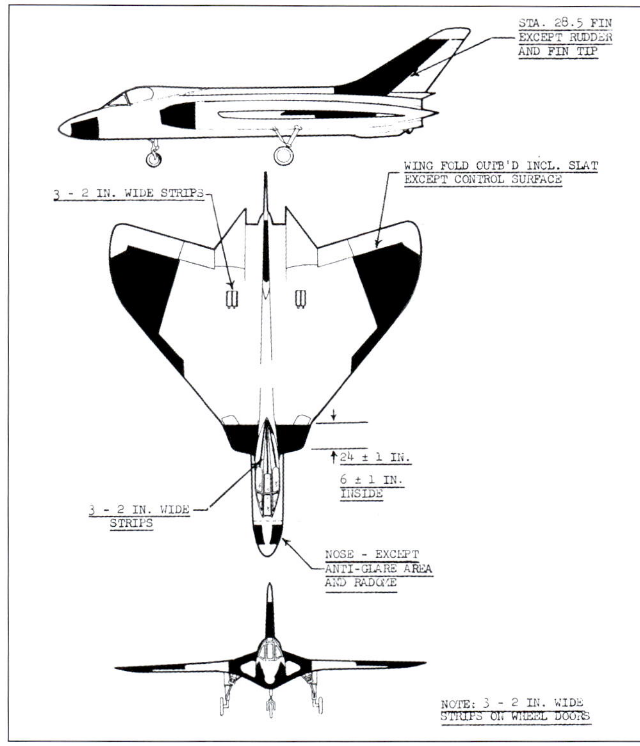

Douglas drawing showing the general markings of the Testing Division Skyrays in the gloss white and insignia red scheme. The markings varied significantly among the aircraft. (Nick Williams Collection)

Various views of the third production Skyray, 130742. The aircraft is carrying rocket pods and drop tanks. BuNo 742 was used for aero demonstrations at Pax River. It flew for 386 hours when it was placed in storage at Litchfield Park, Arizona. (Nick Williams Collection)

for static tests, but flew later for intake duct and autopilot tests. BuNo 743 was used for carrier suitability, but was damaged during a high-sink-rate landing test and was replaced by 747. BuNo 745 was used initially for autopilot and fire control system tests, but was modified later for the successful world time-to-climb record flights. BuNo 748 was used for structural demonstration tests to determine maximum load factors, maximum roll rates, rolling pullouts, and stall and spin characteristics. BuNo 748 was also flown at China Lake for missile launch tests. BuNo 749 replaced 740, which was stricken from damage during a high-g pullout, and 750 logged its entire 808 hours of flight time at NATC in various test programs.[5]

Ship 742 undergoing maintenance at Edwards in 1957. The removed radome reveals the flight recorder used in the Testing Division aircraft. (Nick Williams Collection)

The fourth production Skyray, 130743, at El Segundo. This aircraft flew in a Sea Blue scheme and was used in carrier suitability tests. The deflectors in front of the main gear tires were used for barrier protection. (Nick Williams Collection)

BuNo 743 in flight while it was used for carrier suitability tests at Pax River. This Skyray accumulated 817 hours of flight time before it was stricken from inventory, making it the highest time aircraft of the first 12. Below: A front view of BuNo 743 showing the barrier guards in front of the main gear tires. The early engine air inlets are also evident. (Nick Williams Collection)

The fifth production Skyray, 130744, is shown here at Edwards AFB, but was used as an aero demonstration aircraft at Pax River. This F4D-1 accumulated only 149 hours of flight time when it was stricken from inventory due to accident damage. (Nick Williams Collection)

Even though the first 12 Skyrays were production aircraft, they were essentially hand-built airframes, and were far from standard. Each carried extensive test instrumentation, which varied according to the test mission. All carried flight recorders and cameras in the nose in place of radar. Oscillographs were installed in the electronics bay behind the nose wheel, and a high-speed recorder (a Douglas developed telemetry system) allowed data to be transmitted to the ground. This early telemetry system had limited reliability because of its mechanical switches and vacuum tubes but it was a significant advancement for its time. Some of the Testing Division Skyrays carried movie cameras mounted in streamlined pods on each wingtip and under the belly near the nose wheel well.

On 25 February 1955, eight months after the first flight of the first production F4D, BuNo 130740, Bob Rahn was flying a series of control tests at maximum speed near sea level over the Pacific Ocean. He flew at Mach .98 (750 mph indicated airspeed at 100 feet above sea level) in full afterburner. At this speed the Skyray exhibited a strong Mach tuck from the supersonic airflow over the wing. Full trimmer deflection was required to maintain level flight in a trimmed condition (no force on the control stick). Rahn was satisfied with the test, so he shut off the afterburner and decelerated to subsonic speed. As the Skyray lost speed the tuck-under force disappeared and the trimmer effectiveness increased. The Skyray experienced a 9.1-g pitch-up, exceeding its structural design limit by more than 2 g. Rahn blacked-out from the violence of the force and regained consciousness in a vertical dive heading toward the Pacific Ocean.

He remembers, "I nervously examined the wings of my bird. They were wrinkled from wingtip to wingtip resembling dried prunes. I was in a ruined airplane, but one that was still flying."[6] Upon landing it was discovered that 740 was stressed beyond repair. Not only was the skin severely wrinkled, but also the underlying structure was bent, and the J57 engine had been ripped from its mounts and was lying on the engine compartment access door. The flight recorder provided excellent data on the incident. It was remarkable how much damage was sustained and the airplane was still flyable because it came within 1.4 g of the ultimate design limit, a point at which the Skyray would theoretically break apart. "The aircraft was a 'strike' and not repairable, but we certainly learned a lot from it. Navy pilots were instructed not to trim the airplane in the transonic region at high indicated airspeed."[7]

Reflecting on the mishap Rahn said, "Could what happened on that flight have been predicted? I had logged many sorties at 30,000 feet where the trimmers were deflected at transonic speeds to obtain data for the design of the trim-change compensator, but the pitch-up when rapidly decelerating at high altitude was only 2.0 g. I had flown many times on the deck during the speed runs, but the trimmers were not moved. Here were the clues: The speed runs were made at high temperature with the J40 engine, which had 40 percent less thrust, thus a lower Mach number and less tuck-under, plus the trimmers were not moved.

"On this particular flight the temperature was colder over the ocean and with 40 percent more thrust,

The most colorful Testing Division Skyray was 130748, the ninth production aircraft, shown here taking off from LAX early in its career. Initially ship 748 was used in structural integrity tests. (Nick Williams Collection)

the airplane had accelerated to a higher Mach number than ever before on the deck. This Mach produced the peak tuck-under effect in the transonic region. Coupled with the maximum deflected trimmers, this set up the scenario for me."[8]

In the aftermath of the pitch-up incident, Douglas implemented a Flying Quality Improvement Program. The program produced a solution to the transonic trim problem with the development of an automatic trim-change compensator. This system was designed to automatically provide the necessary amount of pitch trim to overcome Mach tuck while eliminating pitch-up during decelerations. It was a very sophisticated system for the mid 1950s and it required careful adjustments, but when it worked it dramatically improved the Skyray's high-speed handling.

Douglas Test Division pilot Jim Stegman remembers the development of the trim-change compensator, "After we got the trim-change compensator in the airplane we had to go out and adjust these in flight. We had a little wrench, the trim-change compensator wrench, to adjust this. It would throw the pitch trim up at the right Mach number to compensate for the pitch down, or pitch up when decelerating, while in the transonic region. It was an automatic system, but it had to be adjusted and set, so we would always have to go supersonic." During one of these flights Stegman may have achieved a record speed for a production Skyray, "I've gotten pretty close to Mach 1.5 in the F4D-1. During autopilot testing I got special permission to go to 50,000 feet in a clean airplane [no wing tanks]. The dive was a little steeper than normal."[9]

Later in its career BuNo 748 was attached to the Naval Ordnance Test Station (NOTS) at China Lake, California. It was assigned to Douglas test pilot Pete Colapietro who designed new markings to make the Skyray more conspicuous. The original Sea Blue scheme was enhanced with a white-and-red checkerboard on the spine and fin. The radome was light tan, and the wing root bands were DayGlo. The drop tanks and missile pods were white. During the missile launch tests at China Lake, 748 carried movie camera pods on its wingtips and belly. (Nick Williams Collection)

The violent loss of 740 was not the last of Bob Rahn's adventures in the early-production Fords. "Another task was to determine the safe jettison speed at various angles-of-attack for the fuel tanks mounted on pylons under the F4D's wings. This was required because a Skyray pilot on combat patrol might have to quickly get rid of his tanks to enhance maneuverability, rate-of-climb, and maximum speed should he encounter enemy aircraft.

"There was no trouble dropping the tanks at high speeds. Gradually, I worked down toward the slower speeds and with higher angles-of-attack. There might also be an occasion during an approach, for instance, when the pilot had to jettison the tanks because he was jumped by an enemy aircraft.

"Everything appeared normal as the Skyray was cocked up in the landing attitude. I pickled the tank and to my shock and amazement the tank actually flew up and over the leading edge of the wing! When I returned to the base and explained this phenomenon, we concluded that the tanks had developed sufficient lift to fly on their own. Thus they flew up and over the wing! The corrective solution was to avoid jettisoning tanks at slow approach speeds."[10]

Time-to-Climb Records

It was clear from the outset of the flight test program that the Skyray had exceptional climb performance. Even during the carrier suitability trials in 1953 the Skyray displayed an astonishing rate-of-climb. After a launch from the catapult of the USS *Coral Sea* Jim Verdin flew the XF4D-1 to 40,000 feet in less than 5 minutes.

In early 1955 a series of unofficial climb records were being claimed by various Navy aircraft to demonstrate the Fleet's increasing air might. These climbs were timed from brake release to 10,000 feet. The McDonnell F3H Demon set the best time of 71 seconds in February. Bob Rahn recalled that at that time he was flying routine stability and control tests with the F4D. "Late one Friday afternoon, a week after the Demon flight, Al Carder [the F4D Project Coordinator] and I were sitting in his office at Edwards, discussing the next day's test schedule when the subject of the various time-to-climb flights came up. Somewhat puzzled, we both wondered what the other companies were trying to prove, since the Skyray, whatever its other failures, could surely take any climbing contest. After some discussion we decided to see what the Skyray could do the next day 'just for the hell of it,' without any special preparation to the aircraft we were testing and, of course, without any practice.

"At dawn, 19 February, I streaked down the runway at Edwards and pulled the F4D into a near-vertical climb, reaching 10,000 feet in just 56 seconds! My Ford Thunderbird accelerating up to 60 mph in 10 seconds was terrific, but acceleration of the Skyray was astonishing, as I reached 160 mph in the same 10 seconds. I was so busy during the remaining 46 seconds after takeoff, trying to fly a prescribed airspeed and altitude schedule on instruments, that I did not have time to react until the climb terminated. Looking back over my shoulder I saw the runway almost directly underneath me, 10,000 feet away. I was in a 70-degree angle climb traveling at the climb rate of 28,000 fpm."[11]

Rahn's flight was unofficial, but it became highly publicized and it was obvious to many that the Skyray could have easily broken the world time-to-climb record in 1955. However, the record attempt was not to occur for three more years. In February 1958 Douglas

The tenth production Skyray, BuNo 130749, was used to replace BuNo 130740, the aircraft that suffered the 9.1-g pitch-up. (Nick Williams Collection)

The twelfth airframe of the first production block, BuNo 130751, never flew; it was used for static testing. (Nick Williams Collection)

Instrument panels of the Testing Division Skyrays were significantly different from the later production aircraft. Test instrumentation switches were mounted where the radarscope would be placed. (Nick Williams Collection)

requested permission from the Navy to attempt the world time-to-climb records for 3,000, 6,000, 9,000, and 12,000 meters using one of the Test Division Skyrays. The existing record had been set by the French using a Nord 1405 in 1957, but Douglas data suggested that the F4D-1 could break the French records by a substantial margin.

The Navy authorized funds for the record attempt and also approved an attempt for a new record to 15,000 meters, nearly 50,000 feet. Marine Maj. Edward N. LeFaivre, the Skyray Project Officer, was selected as the pilot for the record attempts, and Bob Rahn was named Deputy Director for the record attempt. On 12 May 1958 the planning started with a Douglas briefing for

Bob Rahn generates a visible shock wave during an early morning airspeed calibration pass at Edwards Air Force Base. (Nick Williams Collection)

The second production block provided 12 more Skyrays for the Testing Division. These are two views of BuNo 134760 with unusual drop-tank markings. This aircraft was bailed to the Douglas Testing Division from 12 December 1955 until 28 June 1958. (Nick Williams Collection)

BuNo 134748 was the second aircraft from the second production batch to be used by the Testing Division. It served as a test aircraft from 4 June 1955 until 22 June 1957. The wingtip camera pods are apparent in this photograph. (Nick Williams Collection)

The sixth production aircraft, 130745, at the Douglas Testing Facility at El Segundo on 18 May 1957. Initially 745 was used for autopilot and fire control system evaluations. It was also employed in development of the Sidewinder air-to-air missile. In February 1958 BuNo 130745 was selected as the aircraft for the world time-to-climb record trials. The aircraft was also modified by the elimination of guns, the removal of the tail hook and tail bumper, and the removal of various electronic systems that were not essential for the record flights. (Nick Williams Collection)

Maj. LeFaivre on the techniques needed for the trial flights. The optimum climb angles, the maximum tailpipe temperatures, and the necessary g forces at rotation were defined.

The aircraft assigned to the climb trials was BuNo 130745, the sixth production airframe. It was not exactly a standard-production aircraft since it lacked the 20mm cannon, wing and belly pylons, various electronics, the tail hook, and tail wheel assembly. BuNo 745 also received a specially prepared J57-P-8 engine that could withstand higher operating temperatures and provide increased afterburner thrust. The Naval Air Missile Test Center, Point Mugu, California, was selected as the site for the attempts because of its cool, dense sea air that would maximize thrust and enhance climb performance.

Practice flights were made at Edwards in mid May to familiarize LeFaivre with the climb profiles for each altitude attempt. The 3,000-meter profile required that

BuNo 745 in preparation for the world record time-to-climb trials. All wing store pylons have been removed, the guns have been removed and the ports have been sealed, the tail hook and tail bumper have been removed, and an uprated version of the J57 engine has been installed. (Nick Williams Collection)

Early on the morning of 22 May 1958, Maj. Edward LeFaivre placed 130745 in a 70-degree climb for the 3,000-meter climb record. Here, Maj. LeFaivre poses by the tail of 130745 after setting five world time-to-climb records. Even though the record flights took him to an altitude of 50,000 feet, LeFaivre did not wear a pressure suit. (Nick Williams Collection)

Maj. LeFaivre (center) confers with official timers during his series of time-to-climb record flights. The afterburner exhaust nozzle on the record-setting aircraft was the forged-flap type used with later variants of the Pratt & Whitney J57 turbojet engine, and played a part in the jet's performance. (Nick Williams Collection)

the Skyray lift off at 150 knots, pull 1.4 g until a 70-degree climb was established, and then hold that climb angle to 3,000 meters. Each succeeding altitude trial required an increasingly difficult profile. As the altitudes increased, the engine temperatures increased, and a specified Mach number had to be maintained to ensure the best rate-of-climb. On the 15,000-meter trial LeFaivre was faced with afterburner flameout followed by a core engine flameout above 43,000 feet. He was able to zoom without power to 50,000 feet but his airspeed at that point was 70 knots. He allowed the Skyray to fall forward in a dive to 25,000 feet where he was able to relight the engine.

On 19 May LeFaivre flew 745 to Point Mugu where he continued to practice the climb profiles. He found that the cooler, denser air improved the Skyray's acceleration to such an extent that when he retracted the landing gear the doors were held open by the high-velocity

The Skyray production line early in the 230-airplane order. These aircraft have the original triangular engine inlet with no splitter plate. (Nick Williams Collection)

The Skyray production line at Torrance, California, after the new inlet design was adopted. (Nick Williams Collection)

The Skyray production line during the construction of the final 178 aircraft. By now the production standard was well-defined. (Nick Williams Collection)

Ship 134752 flying with the production standard inlet ducts. The new duct skin has not been painted yet. A NAV-PAC (navigational package) is mounted on the centerline pylon. (Nick Williams Collection)

An early example of the second batch, 134752, as it appeared during production test flight. At this time the aircraft still has the original inlet ducts with no splitter plate. (Nick Williams Collection)

airstream. Fortunately the Douglas technicians were able to increase the pressure of the hydraulic system, allowing the doors to close earlier with more force.

At dawn on 22 May 1958, Maj. LeFaivre flew the first of five official record climb flights. He lifted off at 150 knots and climbed almost vertically in full afterburner on a tail of flame. He reached 3,000 meters in 44.39 seconds—nearly 7 seconds better than the French record. For the remainder of the day, he broke the 6,000- and 9,000-meter records by substantial margins. The next morning he exceeded the 12,000-meter mark by nearly 27 seconds. On the final flight he climbed to 15,000 meters (49,212.5 feet) in 2 minutes 36.05 seconds for a new category of time-to-climb records. The Skyray had now established seven world records—two for speed and five for climb performance.

Volume Production

On 25 April 1952 Douglas received a second letter of intent for 230 additional Skyrays to be built after the initial 12 production articles. Production at the El Segundo plant was not possible. Douglas had developed a large backlog of orders as a result of the Korean War, and several new models, such as the A3D Skywarrior, consumed all of the floor space at El Segundo. The Navy offered Douglas an unused government aluminum plant in Torrance, California, for Skyray production. This facility would provide adequate room for the production lines, but it would require that partially completed airframes be trucked to El Segundo for final assembly and test flight. Douglas agreed to the Navy's offer, and set up a production line for volume production at Torrance. The second production batch was assigned BuNos 134744 to 134973. Eighteen months later in October 1953, after the Skyray established the world speed record, the Navy ordered a third batch of 178 F4Ds, which were assigned BuNos 139030 to 139207.

The production line was frequently interrupted for retrofits and design updates as numerous powerplant, control system, electronic, and aerodynamic problems were solved. One of the most protracted problems was the development of an air-inlet system that would allow the J57 engine to operate properly at high altitude. The problem was particularly capricious.

The Navy's Preliminary Evaluation of production Skyrays noted concern over engine compressor stalls at altitude. Douglas' first attempt to revise the inlet ducts was on 18 March 1955 when 130742 was able to reach 52,200 feet with no sign of compressor stall. But two days later 130742 reached 57,300 feet with the original inlet ducts. There seemed to be no rhyme or reason for the compressor stall problem and many solutions were evaluated. On 15 April 1955 Douglas started an Accelerated Engine Stall Test Program, and by 5 August a definitive inlet duct was decided upon. This consisted of employing a boundary layer ramp with a bleed vent to discharge the turbulent air near the intakes. The intake

A head-on view showing the definitive inlet duct design. The area was increased from the earlier triangular shape. The inlet lips were thickened and rounded. A boundary-layer ramp with bleed vents was attached to each side of the fuselage to prevent turbulent air from entering the inlet duct. (Boeing via Nick Williams Collection)

A November 1955 walk around study of 134749, the sixth article from the second production batch. This Skyray has the finalized inlets with splitter plates. The bulbous store on the right pylon is a portable starting unit. (Nick Williams Collection)

This view of 134749 shows the Iris type of afterburner nozzle of earlier J57 engines. This nozzle had a tendency to operate asymmetrically if some of the "petals" stuck. If this occurred, the afterburner could cause a strong vectored thrust resulting in an unexpected yaw. This view also shows the "half-doghouse" fairing that helped solve the rudder buzz problem. (Boeing via Nick Williams Collection)

A top view of BuNo 134765 as it appeared during production test flight. (Nick Williams Collection)

An F4D-1 during production test flight carrying a full load of rocket pods, drop tanks, and a NAV-PAC. (Nick Williams Collection)

lips were rounded and the intake area was expanded. This decision required that all completed aircraft be retrofitted with the new inlet duct.

Another persistent problem was rudder buffet at high Mach numbers. Like the inlet problem, rudder buffet seemed to defy solution. Numerous attempts to redesign the tail cone were tested. Finally, a new rudder fairing called the "half-doghouse" was decided upon in November 1956. It not only reduced the rudder buzz, but it reduced rudder forces, improved stability, and actually increased top speed. Again, the solution had to be retrofitted to all airframes.

On 28 October 1957 the first aircraft of the final batch was delivered. Fourteen months later on 22 December 1958 the 178th Skyray of that batch, BuNo 139207, was delivered and Skyray production was terminated after a total of 421 of the unique bat-wing aircraft had been built.

A group of recently completed Skyrays awaits production test flights at El Segundo. (Nick Williams Collection)

The final production example, BuNo 139027, on the ramp at El Segundo prior to delivery, 22 December 1958. (Nick Williams Collection)

Early F-102A Production

The XF-92A was an aerodynamic demonstrator (a flying mock-up); only one article was built. However, an airplane that was intended for production, the YF-102, followed it. Convair designed the YF-102 in response to a 1951 Air Force contract for an all-weather Mach 1 interceptor with a ceiling of 50,000 feet. This requirement was very similar to BuAer's definition of the F4D mission.

The YF-102 was essentially scaled up from the XF-92A. It employed the same 60-degree swept delta wing but the wingspan grew to 37 feet from 31 feet 3 inches. The fuselage length grew to 52 feet 6 inches from 42 feet 5 inches. The circular nose air inlet of the XF-92A was replaced by a radome. The engine received air from a pair of shoulder-mounted ducts located on either side of the cockpit. The proposed powerplant was the 21,500-lbf Wright J67, a license-built version of the British Bristol Olympus, but delays in development forced Convair to install the 14,500-lbf Pratt & Whitney J57-P-11, just as delays in the development of the Westinghouse J40-WE-8 forced Douglas to install the Pratt & Whitney J57-P-2 in the production F4Ds.

Convair built the YF-102 under the novel Cook-Craigie plan, a procurement concept that was intended to accelerate the development of new aircraft. The plan eliminated the experimental prototype phase (the XF plane) and moved directly to slow production of service test models (the YF plane). It was thought that flight testing could be conducted on a small number of aircraft that had been fabricated on production tooling, and that full-scale production could be organized simultaneously with the service test program. The Cook-Craigie plan was about to be severely tested.

The YF-102 flew for the first time on 24 October 1953 at Edwards, piloted by Convair test pilot Richard L. Johnson. The initial flight and several succeeding flights were uneventful, but on 1 November the airplane was destroyed in a crash due to an engine malfunction. Johnson was injured, but recovered. Under the Cook-Craigie plan there was a second aircraft ready for flight, and it entered the test program on 11 January 1954. Eight other aircraft were in various stages of completion at the Convair San Diego plant.

As flight testing continued, it became apparent that the YF-102 could not exceed the speed of sound in level flight, a major requirement of the Air Force contract. Cancellation loomed large as Convair scrambled to find a solution. Modifications such as revised air inlets, canopy changes, and lengthening of the forward fuselage were incorporated on the remaining YF-102s. However, the YF-102 top speed would not exceed Mach 0.98.

The YF-102 was intended to be a production aircraft that employed the lessons learned from the XF-92A program. Ten YF-102 aircraft were built before it was determined that the design could not satisfy the Air Force contract requirements. (Convair via Don Spering/AIR Collection)

Two years earlier a NACA scientist, Richard T. Whitcomb, published a research memorandum on transonic drag. NACA made Dr. Whitcomb's paper available to the American aviation industry, but classified it as "Secret." Convair engineers visited NACA's Langley Laboratory to consult with Dr. Whitcomb who applied his theories to the YF-102. It had been previously understood that an aircraft accelerating through the transonic region experiences a rapid rise in drag due to the formation of shock waves. Whitcomb discovered that careful management of the airframe shape could dramatically reduce this transonic drag. His Area Rule theory dictated that the cross-sectional area of the airplane should be kept as constant as possible to mitigate the drag rise. This is especially important where the wing intersects the fuselage since the fuselage cross-sectional area is suddenly increased by the cross-sectional area of the wing. Whitcomb's solution was to reduce the fuselage cross section at wing intersection, giving the fuselage a pinched or wasp-waisted appearance. In the case of the YF-102 it was impossible to constrict the fuselage because of equipment and fuel requirements, so Whitcomb proposed adding area to the fuselage forward and aft of the wing. He reasoned that this would have the same effect as subtracting area at the wing.[12]

It took Convair less than seven months to design and build a new airframe that incorporated Whitcomb's theory. It was designated the YF-102A and featured a fuselage that was lengthened to 63 feet 3.5 inches with a sharpened radome and a V-shaped windshield. It received tail fairings that increased the aft fuselage cross section, and widened cheek inlets to provide additional cross section forward of the wing. The wing was given a cambered leading edge, and its span was increased to 38 feet 1.6 inches. The wing area was now 661.5 square feet, and the empty weight was 19,903 pounds.

On 20 December 1954 Richard Johnson made the first flight of the YF-102A, a short hop to verify the operation of all systems. The next morning Johnson achieved supersonic flight while still in a climb, and was able to reach Mach 1.22 at 53,000 feet. The YF-102A easily met its contract guarantees and validated Whitcomb's Area Rule theory.

The success of the YF-102A made Whitcomb a hero, but it left the Cook-Craigie plan in tatters. The concept of building prototypes with production tooling caused the scrapping of $40 million worth of tools, and left the Air Force with nine barrel-fuselage YF-102s that were virtually worthless. Three more area-rule YF-102As were delivered to Edwards for the Phase I test program, which was now dramatically behind schedule. The first full production aircraft was flown on 24 June 1955, and by January 1956 improvements in the engine inlets allowed it to attain Mach 1.5. In December 1955 the F-102A acquired a larger fin to avoid the violent pitch and yaw divergence condition known as inertia coupling that had destroyed several F-100s. Flight testing was completed in April 1956 and the F-102A entered full service.

The Air Force F-102 program was, from the outset, far larger than the Navy's F4D program. In all, Convair produced 990 examples between 1953 and 1957. The production run consisted of 10 YF-102s, 4 YF-102As, 865 F-102As, and 111 TF-102As (a two-place side-by-side trainer version). All F-102s were produced in Convair's AF Plant No. 2 at San Diego, a facility that built B-24 bombers during World War II. Completed F-102As were flown in primer green to Air Force Plant 42 in Palmdale where they received final equipment and paint. After production test flights, the aircraft were delivered to the Air Force.

The YF-102A employed the area-rule shape suggested by Dr. Richard Whitcomb's research. Also, the fuselage was lengthened, the radome was sharpened, and a V-shaped windshield was installed. The F-102A flew supersonically on its first flight. (Convair via Don Spering/AIR Collection)

The production run included 865 single-place F-102As and 111 TF-102A two-place trainers. (Convair via Mike Machat)

Chapter Five

Flight Characteristics and Systems

The last airworthy Skyray, BuNo. 134806, enters the pattern at NAS Patuxent River after a Test Pilot School familiarization hop in December 1966. In the 15 years following its first flight, the Skyray's control system had been constantly refined to produce crisp, well-harmonized flying qualities. (P. F. Oestreicher)

The final Skyray flight occurred on 25 November 1969 when the last airworthy example, BuNo 134806, was ferried to the National Naval Aviation Museum at Pensacola, Florida. No Skyray has flown in more than 40 years, but its flight behavior can be understood today from the Flight Handbook, training films, personal pilot accounts, and official Naval Air Test Center reports.

Flying the Skyray

The saying goes, "A pilot's favorite airplane is the one he's currently flying." For any pilot fortunate enough to have sat at the controls of a Douglas F4D Skyray, the experience of flying the Navy's first delta-wing jet was indeed memorable.

Normal Flight Procedures

Flight operations began with calculations of takeoff and landing weights to determine runway requirements. The center of gravity was also checked to ensure positive control. A complete exterior inspection of the airplane was commenced at the nose wheel and proceeded around the right fuselage and wing, followed by the tail section and left wing. Items of particular importance included the removal of all engine duct covers to ensure the absence of foreign objects in the ducts, the removal of pitot tube and sensing tube covers, the removal of all landing gear down-lock pins, and the freedom of movement of the slats. The aircraft was boarded, and a complete check of the cockpit was conducted. Important cockpit items included ensuring that the seat ejection safety pins were in place, the master armament switch was off, the stores jettison handle was stowed and safetied, and the engine master switch was off.[1]

Capt. George Watkins (right) delivers 134806 to Capt. John P. Fox (Chief of Naval Air Training, Chief of Staff) and RADM H. S. Mathews (Chief of Naval Air Basic Training) for display at the National Naval Aviation Museum, Pensacola, Florida, at the conclusion of the last Skyray flight on 25 November 1969. (U.S. Navy via Nick Williams Collection)

Engine starts were accomplished with high-pressure air from an external power unit that provided energy to spool up the compressor. The start was accomplished by placing the throttle in the off position, the engine master switch in the on position, the fuel selector switch in the normal position, and depressing the starter button. When the compressor reached 12- to 16-percent RPM the throttle was moved to the ignite position and then to the idle position. The engine would normally start within 20 seconds, as evidenced by a rise in tailpipe temperature and an increase in RPM to 64 to 67 percent. It was important to avoid hot starts (tailpipe temperature exceeds 630 degreesC), aborted starts (engine fails to ignite within 20 seconds), or false starts (engine ignites, but fails to accelerate to idle RPM).[2]

Advancing the throttle to an intermediate setting above idle commenced taxiing. When the aircraft was moving at the desired taxi speed the throttle was reduced to idle. Idle RPM was usually sufficient to maintain taxi speed. Steering was accomplished with differential use of the brakes. The J57-8 engine idled at 68-percent RPM producing about 800 pounds of static thrust, which had to be harnessed with brakes and normal rolling friction. The high idle was used to avoid low-energy compressor stalls and to provide sufficient electrical current to power the fire control system on the ground. Late aircraft used a pop-open nozzle feature, which caused the afterburner nozzle to remain fully open at idle, reducing the thrust produced. Taxiing an early Skyray without the pop-open nozzle required caution to avoid excessive wear and overheating of the brakes.[3]

For airfield takeoffs the pitch trimmers were set at 15-degrees nose up. On the takeoff roll differential braking was used for steering until the rudder became effective at approximately 50 knots. The nose wheel was not to be lifted from the runway until 129 knots was achieved. Lifting the nose wheel any sooner increased drag and lengthened the takeoff run. At 140 knots the aircraft would lift from the runway.[4]

A carrier catapult launch using military thrust required that the trimmers be set for the loading condition and end airspeed anticipated. During the catapult shot the pilot was merely along for the ride. The stick was not gripped during the launch since the pilot could inadvertently apply backpressure causing an over rotation. This was a dangerous condition; aircraft had been lost when over rotated. Slightly less nose-up trim was required for afterburner launches due to the rapid acceleration after leaving the catapult in afterburner. Once the aircraft achieved rotation and commenced its climb, it was often necessary to use forward stick to prevent over rotation. With experience, the pilot could reduce trimmer deflection after launch and fly away with very little stick input.[5]

The landing gear was retracted as soon as a safe landing could no longer be made on a runway. Afterburner takeoffs required that the landing gear be retracted immediately after the wheels left the runway. The retraction cycle could require as much as 12 seconds, and the doors had to close before the air loads would prevent a complete cycle. A significant yaw was often experienced during the retraction cycle since the hydraulic system allowed one main gear leg to lag behind the other. Therefore, it was standard practice to engage the yaw damper before retracting the gear to minimize the effect of the unequal retraction rates. At 206 knots the pitch trimmers would reduce to 6 degrees nose-up automatically. The climb was established at 300 knots. When afterburner was used, the initial climb was not to exceed 500 knots below 10,000 feet to avoid overstressing the aircraft in the transonic range.[6]

Control and stability were always a problem with the Skyray. It had an enormous speed envelope with a stalling speed of 93 knots to a top speed in excess of Mach 1. From its first flight until late in its service life, the Skyray's control system had been constantly modified to produce crisp, well-harmonized responses. Many compromises were made throughout the Skyray's development, which resulted in an airplane that was well behaved at intermediate speeds, but had notable problems at very low or very high airspeeds.

The most frequently mentioned deficiency was the lack of longitudinal stability in the transonic region. The Skyray had a trim-change compensator system that was designed to dampen this instability. When operating properly, the trim-change compensator eliminated most of the nose-down pitch during acceleration, but it didn't adequately compensate for the nose-up pitch during a rapid deceleration. Furthermore, the system was not totally reliable and required frequent adjustment. This pitch instability hindered the Skyray's effectiveness as a gun or rocket-firing platform in the transonic region.

Another control system deficiency was the large amount of adverse yaw that was encountered during roll inputs. Unlike most jets, coordinated turns in the Skyray required a careful application of rudder with small amounts of lateral control. When a large amount of lateral control was used it was difficult to stop the roll within 15 degrees of the desired bank angle.[7]

Stalls were characterized by an airframe buffet that began at approximately 150 knots. The buffet increased in intensity as the airspeed diminished, and was quite pronounced below 120 knots. There was a rapid increase in yawing tendency that could be countered with rudder inputs. Control authority was retained on all three axes up to the stall, but extreme care was required to avoid large angles of sideslip, or a spin would ensue.

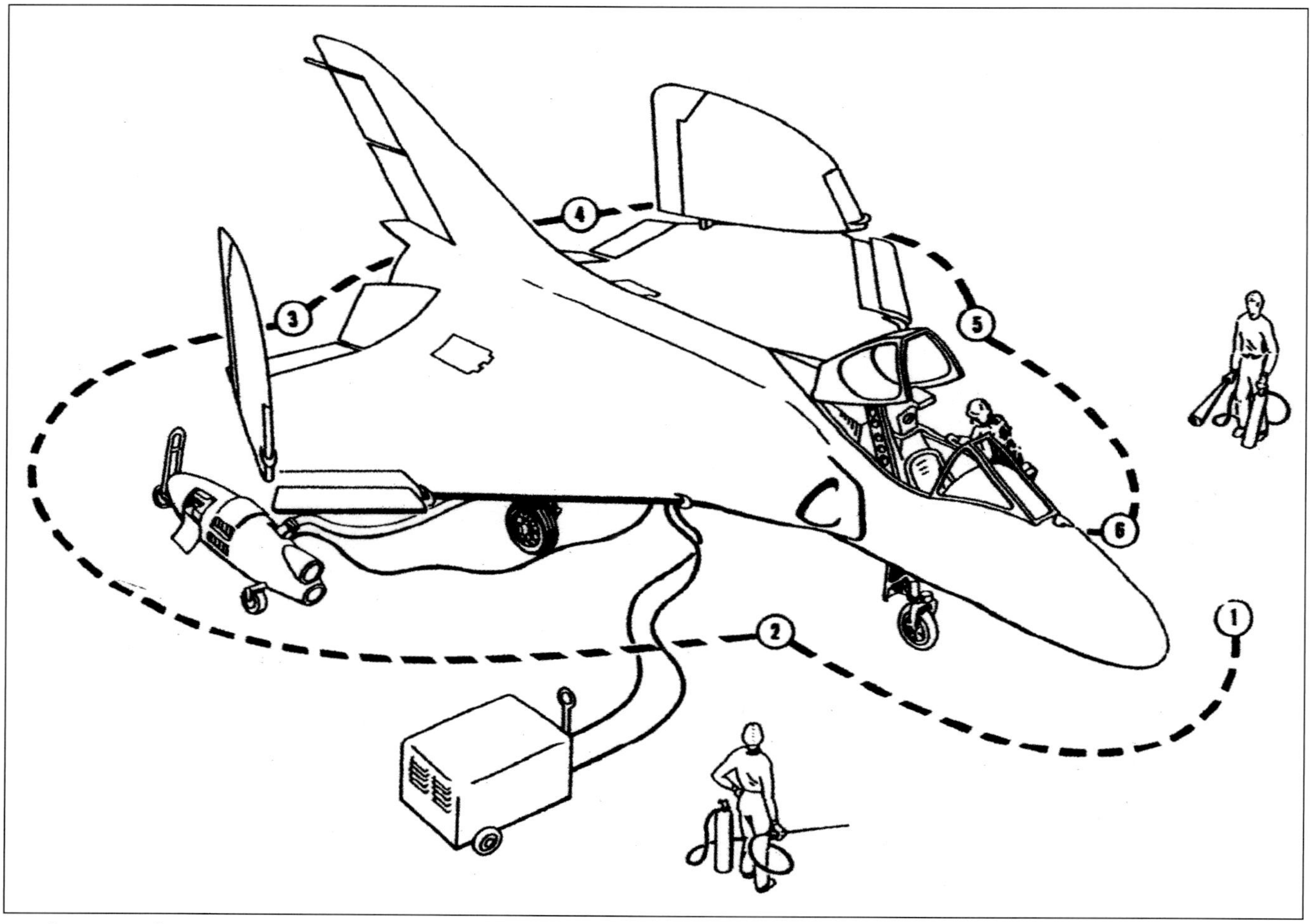

A diagram for the external inspection of a Skyray that was required before flight. (Flight Handbook AN 01-40FBA-1)

Airspeeds below 100 knots could be achieved, but the rate of sink was very high (3,500 to 4,000 fpm) and could not be overcome with military thrust. The actual stall (characterized by a sharp drop of the nose with random wing roll) has occurred as slow as 82 knots, but for practical purposes the Skyray could not be flown below 100 knots.[8]

The roll rate at high airspeeds was phenomenal—more than 500 degrees per second with full stick deflection. Therefore, only small roll inputs were used to avoid over-stressing the airframe or inducing vertigo. The Skyray's low wing loading and ample thrust gave it far better turning capability than any of its contemporaries. Sustained 4-g turns at Mach 0.8 were possible at altitude.

Speed brakes could be deployed at any airspeed. However, they would blowback at airspeeds above 450 knots. They were very effective in reducing airspeed but they could reduce pitch trim at low airspeeds, and were not deployed on final approach.[9] Engine performance at low altitudes was very reliable, and jam accelerations or rapid decelerations of the throttle were tolerated by the J57. But caution had to be exercised above 35,000 feet since compressor stalls and flameouts could occur with rapid throttle movements. The engine relight performance was excellent at altitudes as high as 45,000 feet.

Shore station landings were flown by entering the initial point with 300 knots using 85 to 86 percent RPM. A military break was entered at 1,000 feet above the runway, and speed brakes were opened to decelerate to 185 knots. Landing gear was extended abeam of the touchdown point, and the speed brakes were retracted. Power was maintained throughout the descent and speed was controlled with angle-of-attack. The proper airspeed was 150 knots at the 90-degree point, decreasing to 130 knots on final. The aircraft touched down at 124 knots with 12 to 14 degrees of nose-up trim on the pitch trimmers. Approach speeds for carrier landings were approximately

The pitch trimmers on this VMF(AW)-542 are positioned in the appropriate 15-degree nose-up position for a field takeoff. (Marine Corps via Nick Williams Collection)

Walt Harper, a Douglas Testing Division pilot, was the first person to successfully land a Skyray after a flame-out. The landing was made at NAS Los Alamitos, California. The airframe was written off after the landing. (Nick Williams Collection)

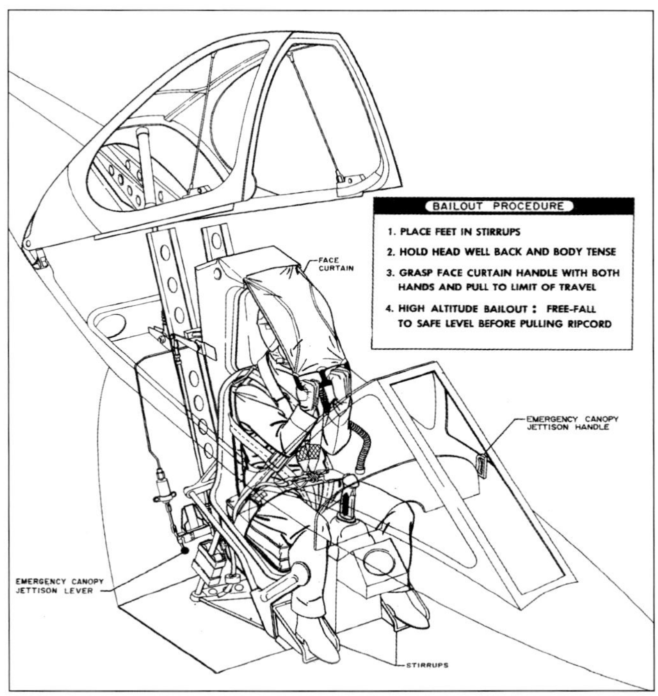

An illustration of the bailout procedure using the Douglas Escapac ejection seat. The Martin-Baker Mk-P5 low altitude ejection seat was used in later Skyrays and retrofitted to many earlier Skyrays. This seat could be used successfully as low as 200 feet. (Flight Handbook AN 01-40FBA-1)

the same as for field landings. Angle-of-attack was of prime importance in carrier landings, and high rates of sink had to be avoided. In a properly flown carrier pattern the distance abeam was about 3,000 yards. The turn off the downwind leg was started abeam of the carrier's stern with a 25-degree bank. At the 90-degree position the turn was gradually flattened until the glide path was entered. The aircraft was held in a constant attitude until contact was made with the deck. The aircraft could not be flared since the tail bumper would contact the deck first, forcing the nose to pitch down, and imposing high loads on the nose gear. Care had to be taken to carry enough power to overcome the turbulence of the carrier's stack wash. The right wing would usually drop as the aircraft approached the round-down area at the forward end of the landing area.

Wave-offs could be safely executed at speeds as low as 115 knots with a typical 17,500-pound gross landing weight. At this speed a small sink rate could be checked with military thrust, and a positive rate-of-climb could be established. The engine required 2.5 seconds to accelerate from approach RPM (85 percent) to military thrust. Normally, the application of afterburner was not necessary for wave-offs, and the momentary loss of thrust during afterburner light-off could be avoided.[10]

Emergency Flight Procedures

Although the J57 enjoyed a reputation for high reliability, engine failures received the most attention in emergency literature. An indication of impending engine failure usually took the form of unstable engine operation. This could be recognized as an erratic increase in tailpipe temperature, a fluctuation in RPM, a compressor pulsation, a loss of thrust and pressure ratio indications, or a lack of response to throttle inputs. Often, retarding the throttle until the engine was "caught" and stable operation was regained could arrest these conditions. If a flameout occurred, successful air starts could be accomplished at altitudes as high as 45,000 feet with an engine windmilling in the range of 12- to 36-percent RPM. A successful air re-light would allow the engine to accelerate to idle RPM within 40 seconds of ignition. It was important to avoid negative g loads during an air start attempt, as proper fuel flow would be interrupted.[11]

If an air start was not possible, the Skyray was an easily controlled glider with a good glide ratio. The best gliding speed was about 220 knots (at a gross weight of 21,000 pounds). A distance of 70 nautical miles could be reached from an altitude of 45,000 feet. If normal hydraulic control was available, a flameout landing could be flown if the initial point could be reached with 8,000 feet of altitude and 170 knots of airspeed. The landing gear was lowered at the initial point. It was recommended that the angle of bank during the letdown not exceed 30 degrees, and that the 180-degree point be reached at 4,000 feet while maintaining 170 knots. The 90-degree point was at 2,500 feet, while still maintaining 170 knots. Final approach was flown at 165 knots with a flare over the runway threshold, touching down at not less than 130 knots. If hydraulic control was lost, a manual control approach could be flown, but it was not recommended. The stick forces were very high in the manual mode and the rate of response was very low. The manual control flameout approach required that the

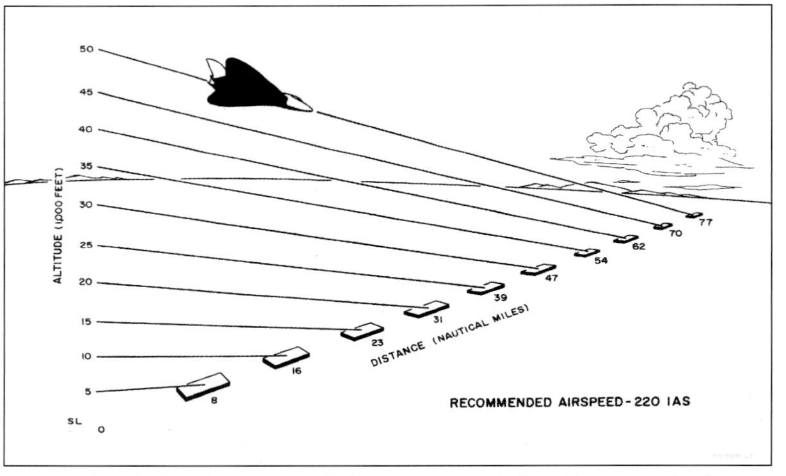

A graph of the flameout gliding range of the Skyray, which was excellent due to its large wing area. (Flight Handbook AN 01-40FBA-1)

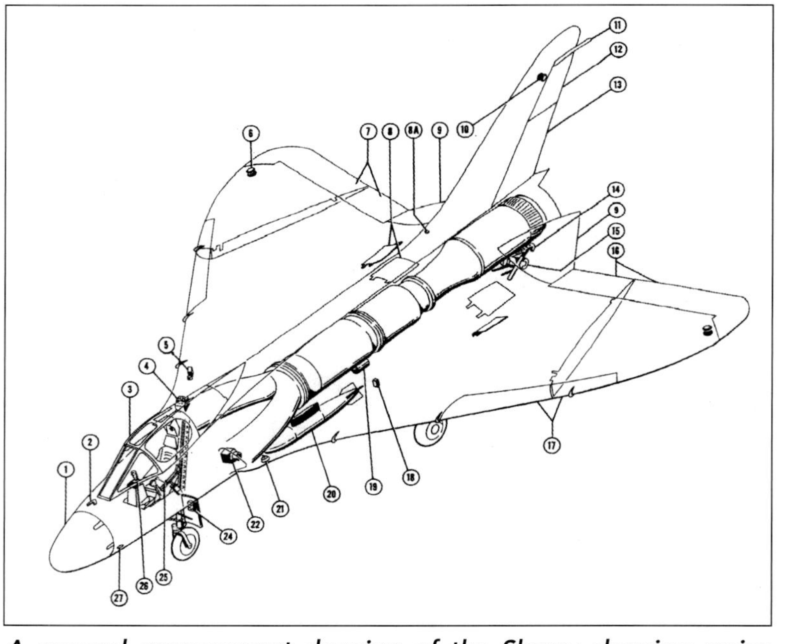

A general arrangement drawing of the Skyray showing major features and systems. (Flight Handbook AN 01-40FBA-1)

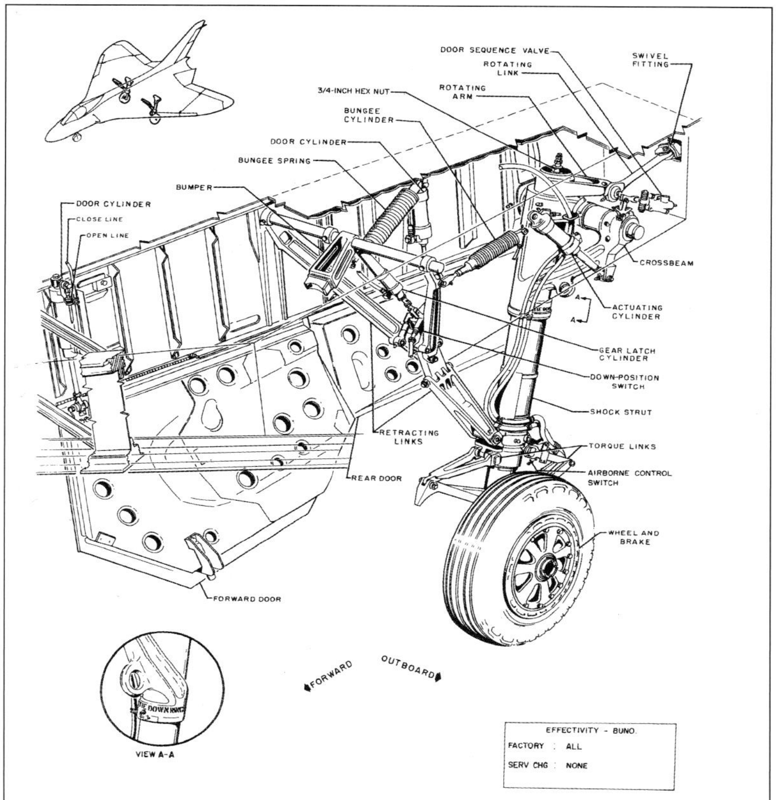

An illustration of the left main landing gear. (NAVWEPS AN 01-40FBA-2)

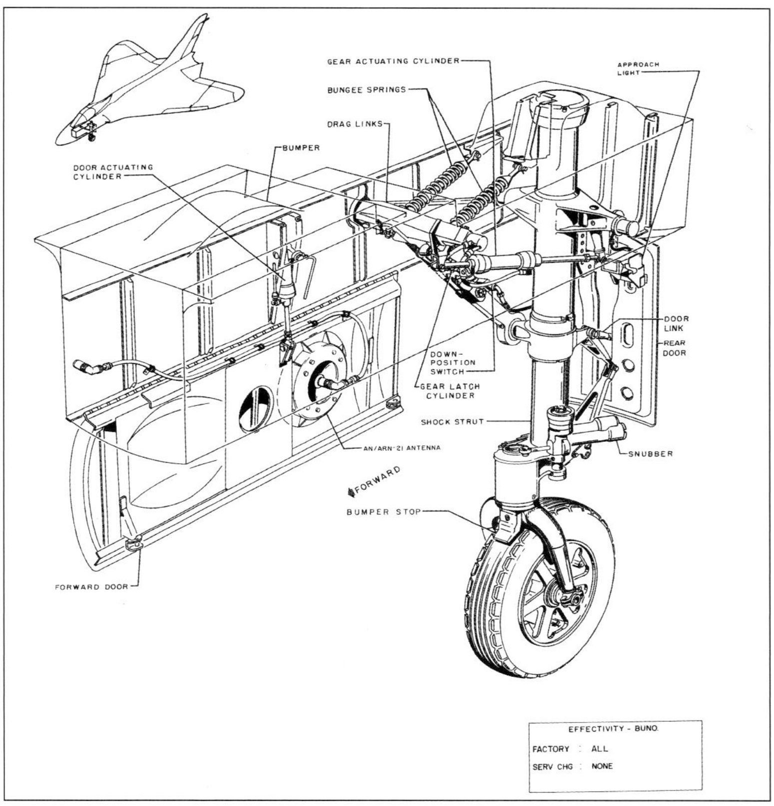

An illustration of the nose landing gear. (NAVWEPS AN 01-40FBA-2)

initial point be reached at 10,000 feet and that the angle of bank did not exceed 20 degrees, and that the pattern was flown with 175 knots of airspeed with touchdown at 140 knots.[12]

Off-runway landings were extremely hazardous, and ejection was considered preferable. If landing on unprepared terrain was attempted, the landing gear should be lowered. The landing gear would shear off upon impact, absorbing much of the energy that would otherwise have been transmitted to the fuselage. Ditching at sea was also extremely hazardous and was not to be attempted if ejection was possible. A ditched Skyray would lose buoyancy in a matter of seconds, and sink. It could become impossible to jettison the canopy below 15 feet of depth on early aircraft. Later models had a pressure relief valve built into the canopy, and canopy jettison was possible to a greater depth.

Early aircraft equipped with the Douglas ejection seat (Escapac) required 800 feet of altitude for a successful ejection. It was recommended that ejections not be attempted at airspeeds above 485 knots to permit adequate clearance from the aircraft. If time permitted, the aircraft should be descended to 10,000 feet and trimmed for straight and level flight. The ejection sequence was initiated by pulling the face curtain handle over the helmet. At that point the canopy would jettison. When the curtain was pulled farther to the pilot's chin the seat would fire. A barometric-opening parachute would deploy automatically at an altitude of 10,000 feet within 3 seconds of seat separation, or it could be opened manually with a D-ring.

Later aircraft were fitted with the Martin-Baker Mk-P5 seat, which permitted ejections from altitudes as low as 50 feet at airspeeds between 120 to 370 knots. The Martin-Baker seat could be used at the Skyray's maximum airspeed with 200 feet of altitude.

If ejection was not possible due to insufficient altitude or airspeed, a ditching could be accomplished by jettisoning all external stores (except fuel tanks if empty). The tail hook was lowered for additional drag and to feel for initial contact with the water. The canopy could be jettisoned at the pilot's discretion prior to contact with the water. The approach was to be made parallel to any uniform swell pattern, and the touchdown should be made on a wave crest. The approach was made in a nose-high attitude. If contact with the water was made in a level attitude the air inlets would scoop water, and cause the aircraft to dive violently.[13]

Landings on rain-soaked runways could also be hazardous. The Skyray's tires were very thin and carried high pressure giving them a tendency to hydroplane on a wet surface. Field landings on rain soaked runways were often flown with the tail hook extended and Morest field arresting gear was used to stop the aircraft. An alternate tactic was to shut down the engine at touchdown, thus eliminating 800 pounds of residual thrust on the rollout. After the landing energy was dissipated the windmilling engine could be re-lighted for taxiing.

Aircraft Systems

Much like the handling and performance of an automobile being dependent on its suspension and steering systems, the handling, performance, and weaponry of a new jet fighter is highly dependent on its flight control, trim, radar, fire control, and armament systems.

Flight Control System

The Skyray's primary flight control system consisted of elevons on the trailing edge of each wing and a rudder on the vertical stabilizer. The elevons served as both elevators and ailerons providing pitch and roll control. The rudder provided yaw control. The elevons were assisted by two pitch trimmers, which provided longitudinal trim for the appropriate flight condition (the trimmers were used most commonly in takeoff, landing, and transonic flight).

The secondary flight control system consisted of four speed brakes and leading-edge slats. All flight controls were actuated from the cockpit except for the slats, which were actuated by aerodynamic forces on the wing. The elevons and speed brakes were driven by one of two redundant 3,000-psi hydraulic systems.

Both systems were self-contained and functioned independently of the other. The primary system was dedicated to powering the elevons and upper rudder. The second (or utility) system was used to provide power for the retractable landing gear, wheel brakes, speed brakes, arresting gear, and wing folding. It also powered the elevons in tandem with the first system. Sufficient hydraulic force was available at subsonic airspeeds to power the elevons if either system failed. In the event of a total hydraulic failure the elevons could be driven manually.

Under manual control the control stick could be lengthened to provide the mechanical advantage necessary to deal with the air loads. The elevon deflection was reduced and enormous manual force was required, but the aircraft could be controlled at moderate airspeeds. The rudder was split into two surfaces with the lower half serving as a cable-driven conventional rudder, and the upper half serving as a servo-driven yaw damper and directional trim device. The pitch trimmers were actuated by an electrically driven jack screw. The primary control system could be governed by an auto pilot known as the D-1 Auto Control system, which provided coordinated control movements by electro-mechanical forces. The D-1 was capable of maintaining level flight, constant barometric altitude, an established compass heading, turns, climbs, dives, and a combination of these maneuvers as selected by the pilot. Elevon sensitivity could be varied by a mechanical-advantage-change system that could reduce the amount of control surface deflection by as much as 2/3 of the normal amount allowing for smoother control at high airspeeds and low altitude.[14]

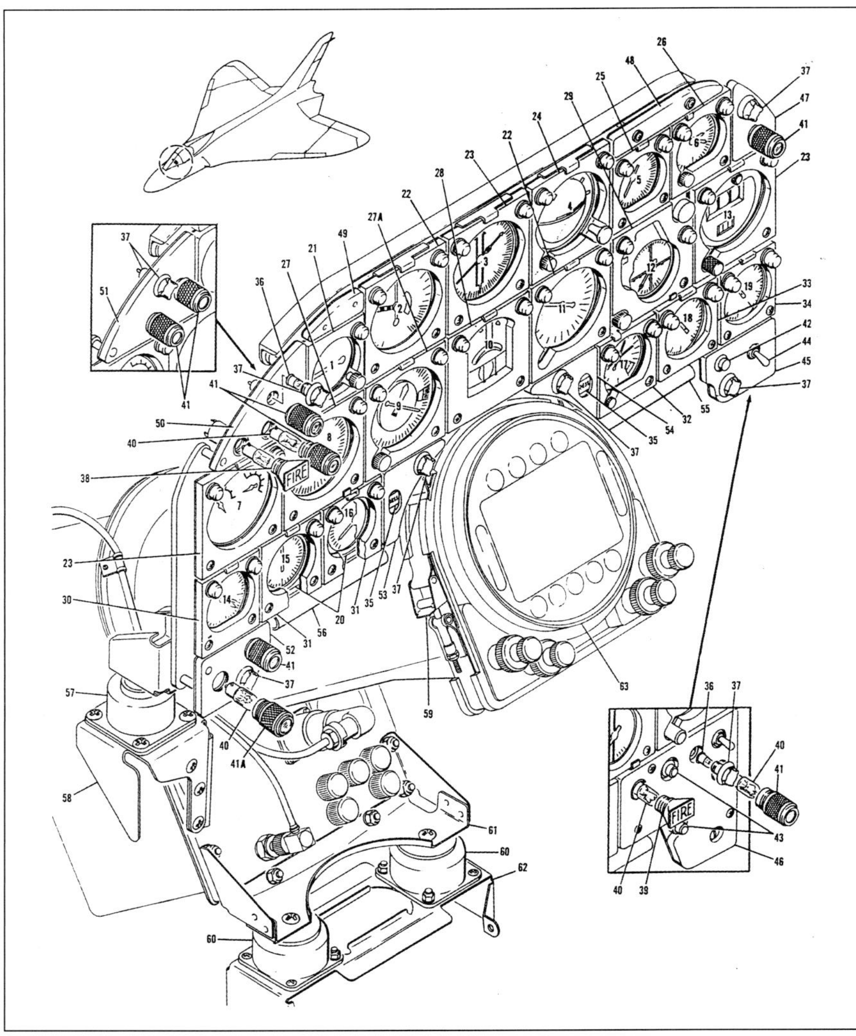

An illustration of the instrument panel showing the location of the radarscope in the lower center. (Parts Handbook AN 01-40FBA-4)

Transonic Trim Change Compensator

This was an electro-mechanical system that automatically trimmed the airplane in the transonic speed range (Mach .9 to 1.2) to compensate for the nose heaviness that occurs near the speed of sound. The trim was obtained by the deflection of the elevons rather than the pitch trimmers. The elevons moved upward as the airplane reached the speed of sound to a maximum deflection of 8.4 degrees and decreased to zero as the airplane decelerated to subsonic flight. The system required in-flight adjustment for each airplane, and pilot accounts mention that frequent readjustment was necessary.

The system functioned by sensing the Mach number and the altitude of the aircraft. As the aircraft approached the speed of sound a Mach transducer sent out a steadily increasing signal until it reached the onset point (approximately Mach .92). The voltage of the signal at this point was sufficient to trigger the system into operation. As the airspeed increased the transducer signal increased, causing a servo actuator to deflect the elevons upward, offsetting the effect of the supersonic shock waves that form above Mach .92. When properly adjusted, the transonic trim change compensator improved the Skyray's high-speed longitudinal stability dramatically. When the system was inoperative, or out of adjustment, low-altitude supersonic flight was avoided.[15]

Landing Gear

The Skyray was equipped with a conventional retractable tricycle landing gear plus a tail bumper gear. The tail bumper gear, which protected the aft fuselage from runway strikes, was extended only for landing. It retracted automatically when weight was on the main gear. All four landing gear were operated by the utility hydraulic system, but they could be extended, in an emergency, by a cable-release system.

The nose gear and main gear retracted forward. The tail bumper gear retracted rearward. The main gear also rotated 90 degrees so that the tire was stowed flat in the well. Two doors covered the nose gear. The forward door was closed when the gear was extended. It opened during the retraction cycle to allow the gear to enter the well, and closed upon completion of the cycle. The opposite sequence occurred upon extension of the gear.

The rear nose wheel door remained open when the gear was extended, and was pulled closed by the nose wheel strut when it retracted. Two doors also covered each main gear. The forward doors remained closed when the gear was extended. They opened during the retraction cycle, and closed on the retracted gear. The rear main gear doors remained open when the gear was extended. In flight the landing gear could require as much as 12 seconds to complete a retraction cycle.[16]

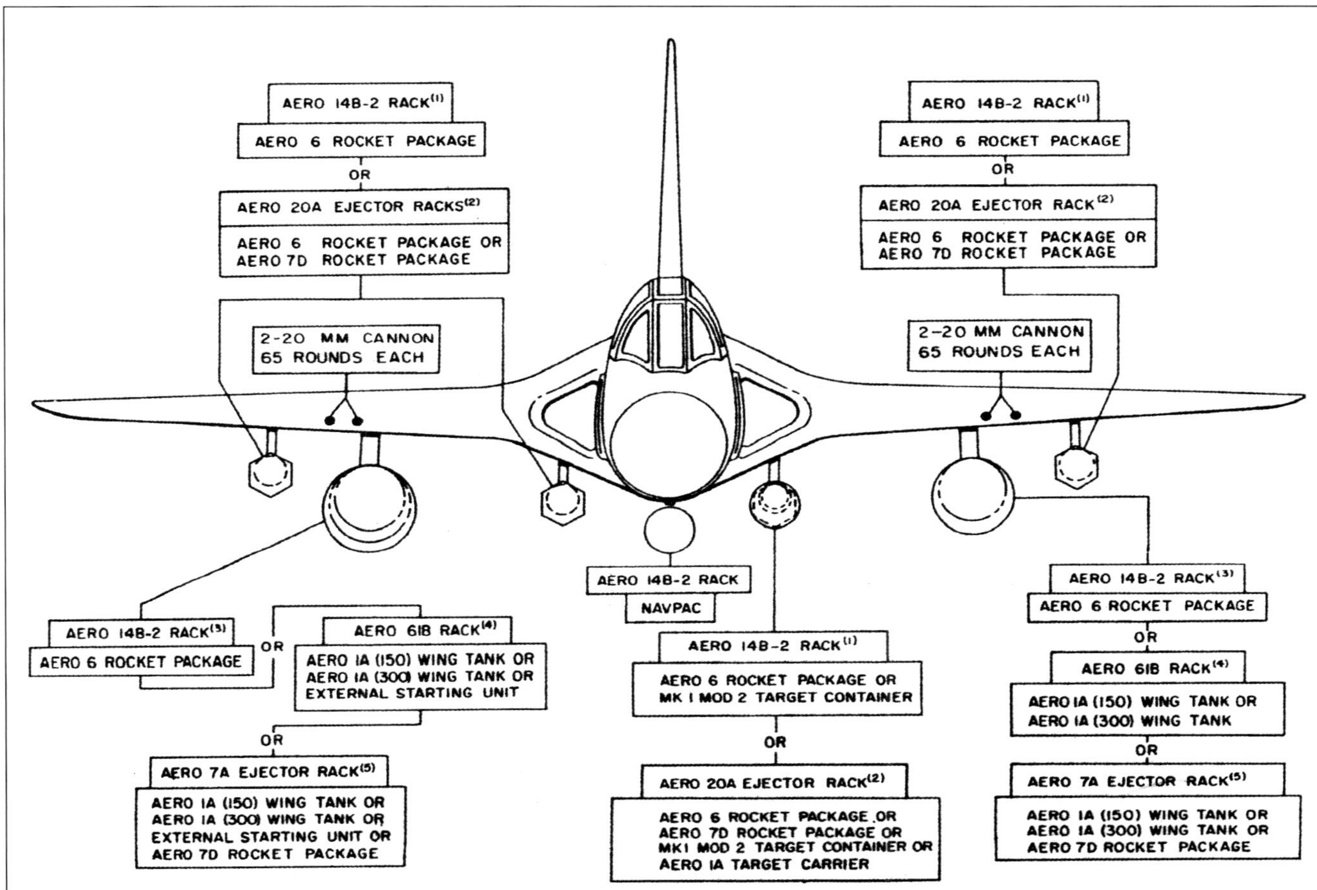

An illustration showing the external stores locations for the Skyray. (Flight Handbook AN 01-40FBA-1)

Radar, Fire Control, and Armament

The radar system used in later production, all-weather F4D-1s was the Westinghouse APQ-50. This was a gyro-stabilized airborne intercept radar that could be used in a search or mapping mode in ranges of 12, 24, 100, or 200 miles. Target acquisition or lock-on could be accomplished in the 12- or 24-mile range only. Once lock-on had occurred, the target was tracked automatically down to a minimum range of 200 yards.

In the attack mode, a circle/dot presentation was displayed to provide steering information and a larger range circle provided range and rate information. The range circle started collapsing at 20 seconds prior to firing in lead pursuit attacks. In lead collision attacks the rocket packages were fired automatically at 500, 1,000, or 1,500 yards depending on the pilot's selection. An Aero 13F Armament Control system provided the firing solution for the intended attack by computing the correct lead angles and steering information.

The Skyray airframe provided seven hard points for various bomb rack and rocket launchers. The aircraft centerline accepted one Aero 14B bomb rack and rocket launcher. Alternatively, this pylon could also carry a specialized navigation and radio package. The right and left wing root positions accepted Aero 20A ejector bomb racks. At 1/3 span an Aero 7A ejector bomb rack was used to carry a 300-gallon drop tank. Finally, at 2/3 span another Aero 20A ejector bomb rack was mounted. The Aero 20A pylons usually carried Aero 6A-1 (7-shot 2.75-inch rockets) or Aero 7D (19-shot 2.75-inch rocket) packages. The Skyray could also carry four Raytheon AIM-9B Sidewinder heat-seeking missiles on special launch rails located at these stations. Four forward electrical impulse firing Mk 12 Mod 0, 20mm guns were mounted in the wing leading edges, two on each side of the aircraft just outboard of the center stores rack. Ammunition was supplied to each gun by a linear chute containing 65 rounds. The Skyray was one of the last Naval fighters to carry wing-mounted guns.[17]

Flying the Delta Dagger

The F4D-1 and the F-102A shared the same tailless delta configuration, both were powered by the Pratt & Whitney J57 engines, both were developed from the work of Dr. Alexander Lippisch, and both entered squadron service in March 1956. They exhibited some similar flight characteristics, but in many ways the F-102A and the F4D-1 were quite different airplanes. The F-102A was much larger with an overall length of 68 feet 2 inches (including the pitot boom), with height of 21 feet 2 inches, and gross weight limit of 32,859 pounds. The Skyray was only 45 feet 8 inches in length, with a height of 13 feet, and a gross weight limit of 27,937 pounds. The F-102A was powered by the J57-P-23A, a significantly higher thrust version of the engine installed in the Skyray.[18]

Normal flight operation of the F-102A began with a calculation of mission fuel requirements, expected landing weight, and center-of-gravity location. The preflight inspection included the necessity of extending the missile bay doors to inspect the internal armament aboard. An external cart was used to provide compressed air and electrical power to start the engine. The F-102A could also be started on stored pneumatic pressure and battery power when no external starter was available. Unlike the Skyray, which used differential braking to taxi, the F-102A used nose wheel steering. The nose wheel could be deflected up to 50 degrees left or right of center, giving the aircraft the ability to make sharp turns. Idle thrust was sufficient for most taxiing operations, but when lower thrust was desired for slippery conditions the idle thrust could be reduced by operating the engine on the emergency fuel system, which provided lower fuel flow and a correspondingly lower idle RPM.[19]

Holding the aircraft in position with the brakes and advancing the throttle to *Full Mil Power* initiated the F-102A takeoff. After a check of the engine pressure ratio gauge (which is a more accurate measure of thrust than RPM on a J57) the brakes were released and the nose gear steering was checked as the aircraft started to roll. Afterburner was selected as the aircraft accelerated; however, the takeoff was aborted if the afterburner failed to light or if it did light but produced a directional change in the takeoff roll. This indicated that the afterburner nozzle had malfunctioned by not opening symmetrically—a condition that produced a vectored thrust making the airplane uncontrollable. Rotation commenced at 125 knots with the nose raised to 15 degrees.

A higher angle would delay the lift-off by increasing drag, and a lower angle would delay lift-off by producing less lift. Landing gear was retracted as soon as possible to avoid exceeding the gear limit speed, and the pitch damper was engaged prior to 320 knots to avoid pilot-induced oscillations.[20]

The F-102A was reputed to have excellent flying qualities. The control harmony and effectiveness from stall to supersonic flight was regarded as exceptionally good. At supersonic speeds the airplane did have a tendency to oscillate in yaw, giving a weaving motion. It also exhibited some Mach tuck or nose heaviness in the transonic range, but not nearly as severe as the Skyray.

An important part of the F-102A preflight inspection was an examination of the missile bay doors and the armament loaded in the bay. Note holes in the lower door for housing smaller unguided air-to-air rockets. (Don Spering/AIR Collection)

Both of these conditions were controlled by yaw and pitch dampers. It was recommended that the F-102A not be flown supersonically with either damper inoperative. Large aileron deflections at high airspeeds were avoided because of the excessive roll rate and the possibility of inertia coupling. Slow-speed flight was very benign with good control effectiveness. A high angle-of-attack was required in slow flight, but pilot visibility was adequate up to 20 degrees. A high sink rate could develop at high angles of attack and low airspeeds without the pilot's awareness and even afterburner thrust might not arrest the rate of descent. Pilots were cautioned to avoid this condition at low altitudes.

The F-102A did not stall in a conventional manner. A stall for this aircraft was defined as loss of control about any one axis. In level flight at approximately 175 knots a slight warning buffet could be detected. A more severe buffet occurred at 130 knots, but the airplane could be flown at 115 knots before encountering a stall. Recovery from a stall was accomplished by releasing backpressure on the stick and acquiring a minimum airspeed of 130 knots. An F-102A weighing 25,000 pounds could lose 6,000 feet of altitude before recovering from a stall. The F-102A tended to experience compressor stalls at altitudes above 35,000 feet when the airspeed dropped below 120 knots. This condition could cause damage to the engine by exceeding the exhaust-gas temperature limit.

Speed brakes could be deployed at any airspeed to provide a rapid deceleration. Extension of the speed brakes would provide a slight pitch-up tendency; conversely, retraction of the speed brakes would provide a slight pitch-down tendency. It was possible for the speed brakes to deploy asymmetrically, which would induce a large yaw. Pilots were advised to be aware of that condition and retract the brakes immediately if it should occur.[21]

A normal landing with 1,500 pounds of fuel remaining commenced with a 300-knot entry to the traffic pattern. Speed brakes were deployed as needed to place the airplane on the downwind leg with 220 knots. Landing gear was extended before turning on the base leg with 180 knots of airspeed. The final approach was flown with 170 knots, with touchdown occurring at 130 knots. The drag chute could be deployed at any speed below 165 knots. It could even be deployed while the F-102A was airborne just before touchdown, but that was not recommended. On touchdown the aircraft was held at a 15-degree angle-of-attack to ensure maximum aerodynamic braking, and the nose wheel was lowered to the runway below 90 knots.

Unlike the Skyray, the F-102A did not have an effective low-altitude ejection system. Ejections below 2,000 feet of altitude, or less than 95 knots of airspeed, were not recommended in the F-102A. However, spin behavior was similar to the Skyray's—difficult and often unpredictable. Intentional spins in the F-102A were prohibited since recovery required an excessive amount of altitude, and there was a substantial probability of exceeding engine temperature limits.

The onset of a spin was characterized by airframe buffet, loss of lateral control, a high sink rate, and engine compressor stalls. If the airspeed was allowed to decrease well below the recommended minimum or a g load was held through a severe buffet the airplane would fall off laterally in either direction, and a spin would develop if stick back-pressure was maintained.

Large aileron deflections at low airspeeds could also induce a spin. The spin rotation would develop in the

An F-102A firing a salvo of AIM-4A Falcon radar-guided air-to-air missiles. (U.S. Air Force via Don Spering/AIR Collection)

opposite direction from the aileron input. Spin recovery required a minimum of 12,000 feet of altitude. If a spin developed below that level the pilot was advised to eject. If the spin occurred above 12,000 feet, the recommended recovery procedure was: Hold the controls neutral, bring the throttle to idle, place the airplane in a clean configuration by jettisoning any external store, retracting the landing gear, closing the armament bay doors, and closing the speed brakes. Once the direction of rotation had been established the aileron was deflected to half of its throw in the direction of the spin.

As the spin breaks, the nose would pitch down and the rotation would stop. The aileron input was then neutralized and smooth back-pressure was applied to pull out of the dive. Spin recovery in the F-102A, as in the F4D-1 Skyray, was not for the faint of heart.[22]

Speed brakes on the F-102A are seen here at the base of the vertical stabilizer. These could be extended at any airspeed, but asymmetrical deployment would produce a violent yaw. (U.S. Air Force via Don Spering/AIR Collection)

Proper takeoff technique required that the F-102A be rotated to a 15-degree angle-of-attack at 125 knots. Held in that attitude, the aircraft would then become airborne at 140 knots. (U.S. Air Force via Don Spering/AIR Collection)

Chapter Six

Skyray Joins the Fleet

BuNo 134777 was the first Skyray delivered to the Fleet. It was accepted by VC-3 at Moffett Field on 16 April 1956, beginning a 13-year career for the F4D-1, which served with more than 35 Navy, Marine, and civilian units. (Boeing via Nick Williams Collection)

By February of 1956 the Navy was convinced that it had a spectacular interceptor in the Skyray. Its performance was remarkable for a carrier-based aircraft with two world speed records and an unofficial time-to-climb record. Many of the Skyray's early quirks had been fixed or tamed; however, there were still serious bugs to be worked out. The J57 engine was new and experiencing some fuel control and afterburner problems. The Aero 13F Armament Control system was delayed and was not installed in the early production aircraft, relegating these Skyrays to the role of day fighter. Yet BuAer was eager to begin Fleet deliveries.

More than seven years had passed since the mock-up was approved—more than five years had passed since the Skyray's first flight. In February 1956 a Fleet Indoctrination Program (FIP) was organized to subject

F4D-1, BuNo 134777, departs from Los Angeles International Airport on the first factory-to-Fleet delivery on 16 April 1956. This Skyray was delivered to VC-3 at NAS Moffett Field, California. (Boeing via Nick Williams Collection)

BuNo 134777, the first Skyray to enter operational service, is parked on the ramp at NAS Moffett Field, California. (Bill Larkins)

several production Skyrays to intense use by Fleet pilots and mechanics in a real-world environment. These aircraft would be employed as if they were in squadron service. Technical manuals would be checked, parts supply would be tested, and maintenance procedures would be verified. If all went well, the Navy expected the first Factory-to-Fleet deliveries to take place in April 1956.

Over the next 13 years the Skyray served with at least 35 separate Navy, Marine, and civilian organizations. It was retired from operational service in February 1964, but several examples continued in service at the U.S. Naval Test Pilot School until November 1969.[1]

Operational Squadrons

Cockpit checkouts, carrier workups, and practice bounces are all part of a Navy pilot's vocabulary. As the Skyray entered operational frontline service, these terms took on new meaning with the Navy's newest jet.

VC-3

Skyray deliveries to Fleet units started on 16 April 1956 when Fleet Composite Squadron Three (VC-3) based at Moffett Field, California, accepted the first aircraft, BuNo 134777, at El Segundo. The Squadron accepted three additional Skyrays by the middle of May 1956. VC-3's mission was to provide Fleet pilot instrument training and indoctrination training for Fleet replacement pilots and maintenance personnel. The markings applied to these early Skyrays at the Douglas factory were quite spartan. They consisted of a dark blue tail band with three yellow stars and a tail modex of NP. Originally, BuNo 134777 wore the nose number 21 when it was delivered, but the Squadron later changed this to 22.

Within three months of the first Skyray delivery, VC-3 was redesignated VF(AW)-3; however, its mission remained unchanged, and the aircraft retained their VC-3 markings. Later the NP tail modex would be replaced with TT, and this Moffett Field squadron would continue to fly their Skyrays for two years until the unit was decommissioned on 2 May 1958.

FAWTUPAC/VF(AW)-3

VF(AW)-3, nicknamed the Blue Nemesis, was probably the most famous squadron to fly the F4D-1 Skyray. The history of this unit can be traced to 1944 when a night operations training unit, NACTUPAC, was formed at Barber's Point, Hawaii. By 1948 this training unit evolved into the Fleet All Weather Training Unit, Pacific (FAWTUPAC), which was assigned the mission of providing

Two additional Skyrays join 134777 at Moffett Field. These aircraft would be used in an intense training program to build experience with the radical new interceptor. (Bill Larkins)

all-weather training for Pacific Fleet pilots. The unit moved to NAS North Island near San Diego, California, in February 1955, and was assigned the supplementary mission of air defense. Because of the dual missions, the squadron functioned as a split unit. Air-defense activities were conducted from North Island while all-weather training was conducted at Moffett Field.

The training unit had been flying Douglas F3D Skyknights, which were augmented with Skyrays in the summer of 1956. On 2 May 1958 the Moffett Field element of VF(AW)-3 was relieved of its all-weather training duties and decommissioned. Simultaneously, the North Island contingent was redesignated VF(AW)-3 and assigned the sole mission of air defense. Adding to the complexity was the fact that the air-defense mission fell under the control of the North American Air Defense Division (NORAD), an Air Force command. VF(AW)-3 was to serve as tactical Naval unit that never went to sea and was under the control of the Air Force—a truly unique organization.

The cream of the F4D community staffed VF(AW)-3, the most experienced Skyray pilots in the Fleet. The air-intercept and defense missions were often conducted with F-101B, F-102A, and F-106 units. CDR Eugene Valencia, a World War II ace, served as Operations Officer, Executive Officer, and finally Commanding Officer. Noted author, Bill Gunston, characterized Valencia as: "...an outstanding ball of fire. One of his precepts was 'The bigger they come, the harder I work 'em,' and he insisted on a flight of three Skyrays being at near readiness every minute of the year. He had a habit of hitting the alert button at about 3 in the morning, and there would be trouble if the 'burners were not blasting on the runway inside 2½ minutes' (2 minutes was expected)."[2]

LT Arland Nankivil reported to VF(AW)-3 in 1959. He had acquired extensive Skyray experience in VF-23, a carrier squadron, before being selected for the elite North Island unit. He recalled a memorable night in 1960 when he scrambled to intercept a "live" target that had penetrated the Air Defense Identification Zone (ADIZ). "The horn sounded and we were immediately cleared to the active [runway]. Once airborne we were vectored to the north by the Air Force radar site. This was not a drill and we were fully armed. When my flight leader and I arrived at 35,000 feet we spotted the target, it was a SAC B-47. We copied his tail number and reported the identification to NORAD. It was determined that the B-47 had strayed more than 100 miles from its planned course. We learned later that this hapless crew was the highest-ranking B-47 crew in SAC. When we returned to North Island the Commanding

Three months after receiving their Skyrays, VC-3 was redesignated VF(AW)-3. Initially the aircraft markings were unchanged, but ultimately the NP tail code was changed to TT. (Bill Larkins)

In August 1956 the markings for the North Island contingent of FAWTUPAC/VF(AW)-3 was this austere scheme of a block-lettered tail code, PA with a nose number. (U.S. Navy via Nick Williams Collection)

A Sidewinder missile is being mounted on this F4D-1 at Moffett Field when VF(AW)-3 still had an all-weather training mission. (Tailhook Association via CDR Doug Siegfried)

An early VF(AW)-3 Skyray stands alert under the squadron's new air-defense role. (Tailhook Association via CDR Doug Siegfried)

CDR Eugene Valencia was the driving force behind VF(AW)-3. He demanded perfection—and got it. VF(AW)-3 outperformed all other Fighter Interceptor Squadrons in NORAD for two consecutive years. (Mark Nankivil Collection)

LT Arland Nankivil and LT Scott Craig pose by a VF(AW)-3 Skyray. This elite squadron was staffed by the cream of the F4D community, and when he was selected for VF(AW)-3, LT Nankivil had completed several cruises with VF-23. (Mark Nankivil Collection)

General of NORAD called to congratulate us on an outstanding job. It was feared that this was a real enemy penetration, and also it was a huge public relations victory to embarrass the Strategic Air Command."[3]

VF(AW)-3 was highly successful in executing its air-defense mission. NORAD conducted an annual competition to determine the best Air Force interceptor squadron, and for two years, 1957 and 1958, CDR Valencia's F4D Navy squadron won the trophy over several highly effective Air Force units, much to the embarrassment of the Air Force. The Navy also held an annual competition called the Naval Air Weapons Meet, in which VF(AW)-3 dominated at Marine Corps Auxiliary Air Station (MCAAS) Yuma in 1959, by achieving both the highest individual (CAPT R. F. Trudeau) and highest team scores with its Skyrays. VF(AW)-3 was also awarded a flight safety award in 1960 and a Navy Unit Commendation in 1963.

The achievements of VF(AW)-3 were notable, but the markings carried on its aircraft were spectacular. The initial scheme applied to FAWTUPAC Skyrays at Moffett Field was subdued. It consisted of a black PA tail modex and nose number. In early 1958 one squadron of aircraft was painted in a special scheme for the

This VF(AW)-3 Skyray was given special markings for the 1958 Naval Air Weapons Meet at NAAS El Centro, California. The tail code PA was black; the trim on the fin, dorsal spine, and drop tanks was red. This was the only VF(AW)-3 Skyray to wear the red scheme. (Boeing via Nick Williams Collection)

The most striking scheme worn by any Skyray unit was this design on VF(AW)-3 aircraft. (Tailhook Association via CDR Doug Siegfried)

A VF(AW)-3 Skyray over Naval Air Station North Island. (Tailhook Association via CDR Doug Siegfried)

A VF(AW)-3 Skyray accelerates in full afterburner. (Tailhook Association via CDR Doug Siegfried)

A VF(AW)-3 Skyray fires a salvo of 2.75-inch unguided rockets. (Tailhook Association via CDR Doug Siegfried)

A VF(AW)-3 Skyray fires two Sidewinder missiles from its inboard pylons. (Tailhook Association via CDR Doug Siegfried)

The VF(AW)-3 flight line at NAS North Island in the late 1950s. (Tailhook Association via CDR Doug Siegfried)

Naval Air Weapons Meet at Naval Auxiliary Air Station (NAAS) El Centro, California. The PA was slanted and painted in black with a red trim pattern on the tail and dorsal spine. The drop tanks also carried a red trim design. Later that year a more flamboyant scheme was applied to other VF(AW)-3 Skyrays. The tail and spine were painted Insignia Blue with a random sprinkling of yellow stars. The PA modex was painted white and slanted to match the sweep of the tail with a large yellow lightning bolt running across the area. The elevon counterbalances were also Insignia Blue with yellow stars, and the white drop tanks received an Insignia Blue nose and

The ground equipment necessary to support a Skyray is illustrated in this picture. (Tailhook Association via CDR Doug Siegfried)

Operations on the VF(AW)-3 flight line in October 1958. (Tailhook Association via CDR Doug Siegfried)

Two VF(AW)-3 Skyrays on patrol. LT Arland Nankivil is piloting the aircraft with nose number 17. (Tailhook Association via CDR Doug Siegfried)

A well marked VF(AW)-3 Skyray on display at an air show in 1958. (Tailhook Association via CDR Doug Siegfried)

A VF(AW)-3 Skyray (the squadron was still designated FAWTUPAC) at the Third Annual Naval Air Weapons Meet at NAAS El Centro, California, on 14 April 1958. (U.S. Navy via Nick Williams Collection)

The Blue Nemesis pilots of VF(AW)-3 were known for their high esprit de corps. They often wore headgear such as these berets to signify the elite nature of the squadron. (Tailhook Association via CDR Doug Siegfried)

tapering stripe with yellow stars. The VF(AW)-3 squadron badge was placed at the rear of the canopy.

In 1961, VF(AW)-3 Skyrays were deployed to NAS Key West, Florida, during the Cuban Missile Crisis. They flew frequent patrol missions but never engaged. The squadron was decommissioned on 4 March 1963, as the Skyray was being replaced by the Crusader and Phantom II.

VMF(AW)-115

Shortly after the first delivery to VC-3, the Silver Eagles of VMF-115 received its first Skyray, 134758, at Marine Corps Air Station (MCAS) El Toro, California. As with the aircraft delivered to VC-3, the initial VMF-115 Skyrays were delivered without radar. Therefore, they could only function in a day-fighter role. Nevertheless the Marine squadron began an intensive program of flight operations to turn the Skyray into a viable fighting machine. Unfortunately, the Skyray's early service with VMF-115 was marred by a series of accidents. Commanding Officer, LtCol J. S. Flickinger, was killed while attempting a dead-stick landing at El Toro. Frequent compressor stalls, fuel transfer system problems, and structural issues plagued these early aircraft.

The F4D training program was delayed as the aircraft were grounded for numerous modifications and service changes. On 31 December 1956 the first radar equipped Skyrays were delivered to VMF-115, and the squadron was redesignated VFM(AW)-115 to reflect its new all-weather role. By August 1957, the modifications and training had progressed sufficiently for the squadron to deploy to MCAAS Mojave for Phase I combat readiness and instrument training. More than 450 sorties totaling 1,028 flight hours were flown by the VMF(AW)-115 pilots, but the accidents persisted. The Skyrays were grounded for a second time in December 1957 for further modifications to the fuel cells, afterburners, and wing pylons.

When cleared for flight operations, the VMF(AW)-115 pilots continued Phase II training in their Skyrays at NAAS El Centro. Air-to-air rocketry, gunnery, and missile practice was accomplished. In February 1958, the squadron was shipped to Japan aboard the USS

In December 1956 VMF-115 received its first radar equipped F4D-1. The squadron was redesignated VMF(AW)-115 to reflect its new all-weather capability. New markings were applied to the aircraft as shown by this Skyray. (Nick Williams Collection)

The new radar-equipped F4D-1s assigned to VMF(AW)-115 wearing their enhanced markings shown in training exercises over MCAS El Toro on 8 April 1957. (Boeing via Nick Williams Collection)

The squadron modex was changed from AE to VE. When the new tail code was applied a second set of stripes was added to the lower portion of the tail. The full BuNo and Marine were painted at the base of the tail. (Nick Williams Collection)

During the USS* Independence *(CVA-62) cruise in 1962 the VMF(AW)-115 Skyrays wore special markings. The tail code was changed to AG, a red chevron was applied to the rudder, and the squadron insignia was applied to the rear of the canopy. (U.S. Navy via Nick Williams Collection)

VMF(AW)-115 Skyrays on the deck of the USS* Independence *(CVA-62) in 1962. Refueling booms had been installed on the left drop tank for this cruise. (Flight International via Nick Williams Collection)

Windham Bay (CVE-92). Upon arrival they were assigned to Marine Air Group 11, 1st Marine Air Wing, stationed at NAS Atsugi, Japan.

It was during this deployment that the Skyrays came close to combat during an international incident known as the Second Taiwan Strait Crisis. On 23 August 1958, the Chinese Communist People's Liberation Army began an intense artillery bombardment of the Nationalist Chinese forces on the island of Quemoy. The United States had a defense treaty with the Nationalist Chinese and came to their aid by ordering the U.S. Navy

This Skyray was declared surplus and stored at NAS Litchfield Park after the USS Independence _(CVA-62) cruise. It is still wearing the markings from the Mediterranean cruise._ (Nick Williams Collection)

to help protect the Quemoy supply lines. As part of this effort, the Navy modified 100 Nationalist Chinese F-86 fighters by installing the new AIM-9 Sidewinder missile system, which proved highly effective in downing Chinese Communist MiG-17s. Skyrays from VMF(AW)-115 were deployed to North Ping Tung, Taiwan, to fly round-the-clock air combat patrol missions in support of Nationalist C-46 transport aircraft. These transport aircraft were the Nationalist Chinese's sole means of supplying the island of Quemoy. Several Skyray pilots reported Sidewinder "lock-ons" with approaching MiG-17s, but the MiGs always retreated to the Chinese mainland. The Skyrays fired no missiles even though the encounters lasted for six weeks.

After the Second Taiwan Strait Crisis, VMF(AW)-115 Skyrays returned to NAS Atsugi. The squadron continued to operate training missions until May 1959 when the personnel were transferred to other units. A replacement

VMF(AW)-115 Skyrays wore this scheme for a brief period in late 1959. The dorsal spine, tail, and drop tanks carried large panels of DayGlo Red-Orange. (Robert O'Dell via Nick Williams Collection)

Toward the end of their career VMF(AW)-115 Skyrays subdued markings. This F4D-1 is making a field arrested landing at MCAS Cherry Point on 9 August 1962. Approximately 18 months later the squadron would transition to F-4B Phantom IIs. (Boeing via Nick Williams Collection)

group, VMF(AW)-531, was assigned to take over the Skyrays at Atsugi.

In April 1962, VMF(AW)-115 replaced a Navy fighter unit aboard the USS *Independence* (CVA-62) as part of a Carrier Air Group Seven Mediterranean cruise. In October 1962, the VMF(AW)-115 Skyrays were engaged in another crisis when they deployed to NAS Guantanamo Bay, Cuba, in support of President Kennedy's blockade during the Cuban Missile Crisis. They returned to MCAS Cherry Point, North Carolina, in late 1962. Within a year, VMF(AW)-115 started replacing the aging Skyrays with the new F-4B Phantom II. On 29 February 1964 the squadron retired its last F4D-1, making VMF(AW)-115 the longest Fleet operator of the Skyray.[4]

VMF(AW)-314

In December 1957, another Marine all-weather squadron, VMF(AW)-314 Black Knights, converted to the F4D-1 Skyray. After six months of intensive training, VMF(AW)-314 joined VMF(AW)-115 as part of Marine Air Group 11 at Atsugi. On 30 September 1958, they were ordered to North Ping Tung to aid the Nationalist Chinese in the defense of Quemoy and Matsu islands.

Maj. James R. Campbell, USMC (Ret.), recalled his first assignment after graduating from flight training, "I was assigned to Marine All Weather Fighter Squadron 314 to become an F4D-1 Skyray pilot. I was elated, as this was the airplane I had hoped to fly. My position was to be a back up in the event a pilot already in the squadron became unable to deploy to Asia in July. The Wing was still getting new F4D aircraft and if I didn't deploy, I was slated to be transferred to one of the squadrons awaiting aircraft. Two months later, one of our pilots was killed in a car accident and I was selected to deploy with the squadron. We left the USA in early July and were in place at NAS Atsugi, Japan, about four weeks later. Since I was not fully qualified to stand alert (I had not even completed familiarization stage yet), I began a rapid training program. Actually, none of our pilots were fully qualified as our aircraft production was behind schedule, but most had some experience to build on.

"On 24 August 1958, our entire air group, MAG-11, was alerted to deploy to Taiwan as the 'Red' Chinese had attacked the islands of Quemoy and Matsu in the Formosa Straits... The resupply of the islands was being conducted at night by Taiwanese C-46s, but without night/all-weather support capability, the slow moving transports were at the mercy of radar-equipped *ChiCom* fighters. Our F4Ds were the only night/all-weather fighter-interceptors available to protect the transports and to protect Taiwan against the event of an attack.

"Flying out of Ping Tung was generally quite pleasant. The area around the field was generally free of obstructions and weather was usually not a factor. While we often had rain, thunderstorms, or cloudy weather, visibility at low takeoff/landing altitudes was quite good. In the event of rare, really bad weather, there were at least a couple of other airfields in the area that we could divert to. About the only serious weather problem we had was high humidity, which sometimes caused reduced visibility by fogging up the insides of the aircraft canopies.

Two VMF(AW)-314 Skyrays on a training mission from MCAS El Toro, California, on 12 May 1958. (U.S. Marine Corps via Nick Williams Collection)

A VMF(AW)-314 Skyray assigned to 1Lt James "Soupy" Campbell on the hot pad at Ping Tung North Field during the Taiwan Strait Crisis in 1958. (James Campbell)

A VMF(AW)-314 Skyray on the hot pad ready for scramble. It is armed with rocket pods on the outboard pylons, and Sidewinder missiles on the inboard pylons. (James Campbell)

1Lt James "Soupy" Campbell performs a preflight inspection on a VMF(AW)-314 Skyray at Ping Tung in late 1958. (James Campbell)

"We stood alert duty daily. During times of bad weather and each night, we had two aircraft on alert and two more with crews on standby. Almost every crew was launched at least once during each watch. While actual launches against an enemy were rare, we often flew practice missions. A typical mission aircraft carried two rocket packs, two Sidewinder missiles, and four armed 20mm cannon.

"When we got a scramble alert, we would run to the airplane and strap in while starting the engine. As soon as we got on the runway, we ran the engine to full power and engaged the afterburner. Climb-out was done at 400 knots or .88 times the speed of sound to our assigned altitude. While

VMF(AW)-314 Skyrays photographed in spring 1962. The markings consisted of a block lettered tail code VW and nose number. Within a year the squadron would relinquish these F4Ds for new F-4B Phantom IIs. (Roger Besecker via Nick Williams Collection)

climbing, we would adjust the radar, arm the rockets, and begin the search for the enemy. Our primary method of attack was a lead collision intercept firing all 14 rockets in a shotgun-like blast. After firing the rockets, we would turn behind the target and use our missiles. Guns were reserved for a last resort. All weapons could be fired using radar. Although we often scrambled, even when scrambled against an actual enemy we never actually engaged their aircraft, as the enemy would turn away when they realized we were coming.

"Looking back, there were some interesting points. This was the first time, I believe, that an air-group-sized organization in the Marine Corps had ever deployed in a potential combat situation solely as an aviation arm; not in support of ground forces."[5]

From 30 September 1958 until 18 March 1959, VMF(AW)-314 flew 152 scrambles in support of the Nationalist Chinese supply efforts. In gratitude, the Nationalist government presented the squadron with a Chinese ceremonial sword, which was used to dub each pilot qualifying in the F4D-1.VMF(AW)-314 returned to Atsugi on 23 March 1959 and remained there until 8 September, when the squadron was cadred and reformed on the same day at MCAS El Toro. At that time the Marine Corps had a practice of leaving assets in place for replacements units to use. So the Skyrays of VMF(AW)-314 remained at Atsugi and were taken over by the replacement unit, VMF(AW)-542.

The newly formed VMF(AW)-314 at El Toro received new F4D-1s and set out for a training deployment at MCAAS Yuma, Arizona. During 1960 the squadron flew numerous ordnance, radar, navigation, instrument, and field carrier training missions at Yuma and Fallon, Nevada.

From 11 to 14 July 1960 the squadron flew 68 sorties during the Operational Readiness Inspection at Yuma. The Wing Inspector found VMF(AW)-314 to be the most qualified F4D squadron from the West Coast.

On 14 October 1960 VMF(AW)-314 left their Skyrays at El Toro and were flown to NAS Atsugi, Japan, to replace VMF(AW)-542. They reacquired the same F4Ds that they relinquished 13 months earlier, and began a period of intensive carrier qualifications, air defense exercises, and deployments to Itazuke AFB, Naval Air Facility (NAF) Naha, and MCAS Iwakuni. VMF(AW)-314 departed Atsugi in November 1961 and returned to El Toro. They flew F4Ds until the fall of 1962 when they transitioned to McDonnell Phantom IIs.[6]

VF-213

The *Black Lions,* based at Moffett Field, California, received their first Skyrays in February 1957. After several months of Phase I instrument and tactics training the squadron deployed to NAAS Fallon for gunnery training in October 1957. The squadron performed very successfully at Fallon, achieving the highest F4D-1 aerial gunnery scores to date. In March 1958 VF-213 deployed to NAAS El Centro for rocket-firing practice. Again the squadron's performance was commendable, and in April they were able to prove their skills by winning the air-to-air competition at the Third Annual Naval Air Weapons Meet. They were also awarded the James V. Forrestal Memorial Trophy, the Navy League Champions Trophy, the Team Championship Trophy, the Individual Champions Trophy, and 52 other trophies. A stunning display for a unit that had been flying F4Ds for little more than one year.

VF-213 returned to NAS Moffett Field after the Naval Air Weapons Meet, and started carrier qualifications aboard the USS *Lexington* (CVA-16). On 15 July 1957 the squadron embarked on a West Pacific cruise aboard the *Lexington*. During this cruise VF-213 Skyrays

VF-213 Skyrays on display at Moffett Field on 15 September 1957. The squadron received its first Skyray in February 1957. (Bill Larkins)

VF-213 on a combat air patrol mission on 3 November 1959. The squadron was on a Pacific cruise aboard the USS* Lexington *(CVA-16) at the time. Notice the checkerboard pattern has been reduced to a narrow band above the tail code. (U.S. Navy via Nick Williams Collection)

A VF-213 in full burner prepares to launch from the USS* Lexington *(CVA-16). (Tailhook Association via CDR Doug Siegfried)

A two-ship formation of VF-213 Skyrays over NAAS El Centro, California, on 14 April 1958 at the start of the Third Annual Naval Air Weapons Meet. (U.S. Navy via Nick Williams Collection)

The markings worn by VF-102 during the Mediterranean cruise aboard the USS **Forrestal** ***(CVA-59) in the fall of 1958. The diamondback scheme was painted on the dorsal spine, and the tail code was AK.*** (Mark Nankivil Collection)

A two-ship formation of VF-102 Skyrays from the USS **Forrestal** ***(CVA-59) in April 1960. The squadron used the tail code AJ as part of Carrier Air Group Eight during this cruise.*** (U.S. Navy via Nick Williams Collection)

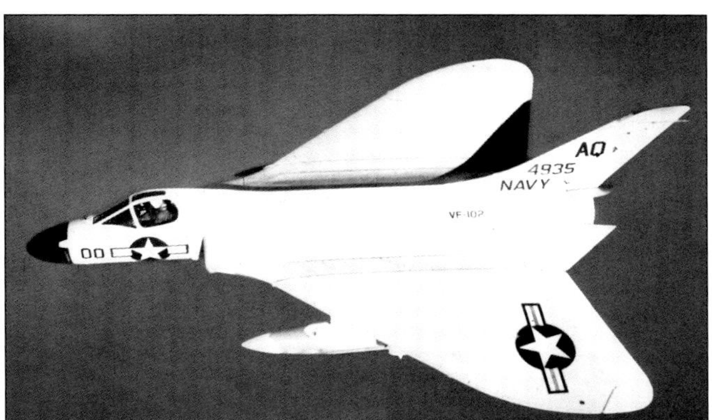

The early markings of VF-102 Skyrays were very simple. The squadron used a tail code of AQ and there were no markings on the upper right wing. (U.S. Navy via Nick Williams Collection)

participated in Operation Blue Skies, a 7th Fleet exercise in the Taiwan Straits, where they provided Combat Air Patrol.

In January 1959 VF-213 flew to NAAS Fallon for additional weapons training, returning to the *Lexington* in March for another series of carrier qualifications. On 25 April 1959 the squadron boarded the *Lexington* for their final WestPac cruise.

In Hawaii VF-213 underwent an Operational Readiness Inspection and received an admirable score of 94.48. From July to December 1959 the WestPac cruise continued and VF-213 participated in three exercises, which included simulated strikes on Taiwan and the southern coast of Japan. During the cruise 14 VF-213 pilots logged more than 1,000 hours, and 10 pilots logged their 100th arrested landing. On 14 December 1959 the

After completing a trap on 7 April 1960 aboard the USS* Forrestal *(CVA-59) the CAG aircraft folds its wings and prepares to park. (U.S. Navy via Nick Williams Collection)

Black Lions of VF-213 began to replace their F4D Skyrays with F3H-2 Demons.

VF-213 Skyrays wore a distinctive scheme of light blue and white checkerboard with the modex NP on the tail. A light blue arrow ran down the dorsal spine from the canopy to the tail, and a light blue design appeared on the drop tanks.[7]

VF-102

The VF-102 *Diamondbacks* based at NAS Cecil Field, Florida, began transitioning to F4D-1 Skyrays from F2H-4 Banshees in the early spring of 1957. They spent the next nine months in training with their new Skyrays while assigned to Air Task Group 202. In early 1958 VF-102 flew aboard the USS *Ranger* (CVA-61) for carrier qualifications. In March 1958 they travelled to NAS Key West for advanced ground controlled intercept training and high-altitude rocket firing. In April 1958 the *Diamondbacks* were selected to represent the Atlantic Fleet in the Navy Weapons Meet at NAAS El Centro where they took second place overall. In the fall of 1958 VF-102 boarded the USS *Forrestal* (CVA-59) for a Mediterranean cruise. At the conclusion of this cruise VF-102 participated in a weapons evaluation exercise named Operation Wexval-9. They boarded the *Forrestal* in January 1960 as part of CVG-8 for a second Med cruise. A squadron operations record was set in June 1960 when the *Diamondback* Skyrays flew 538 hours.

On 9 February 1961 VF-102 boarded the *Forrestal* for their final Mediterranean cruise. Another squadron operations record was set when the *Diamondback* pilots flew 540 hours in 15 days, but several accidents marred the record. The F4D airframes were beginning to show their age, and at least one VF-102 Skyray had its lower fuselage ripped out when a tail hook assembly failed during a trap. VF-102 returned from the cruise in August 1961 and began to prepare for the transition to the new McDonnell Phantom II.

VF-102 Skyrays wore a bright red and white diamondback scheme on their dorsal spine, fin, and drop tanks. Initially, in 1957, VF-102 Skyrays carried a modex of AQ on their fin. This was changed to AK in late 1958, and became AJ in 1960.[8]

VF-74

The VF-74 *Be-Devilers* flew the F4D-1 Skyray longer than any other Navy fleet squadron. The squadron's first F4D-1 was accepted at NAS Oceana, Virginia, on 17 April 1956. A small group of officers that had participated in the Skyray Fleet Indoctrination Program in February 1956 formed the nucleus of VF-74. Initially, the aircraft markings consisted of a tail code C

This Skyray was delivered to VF-74 at NAS Oceana on 17 April 1956. At that time the squadron markings were very plain. A tail code of C was used. The blunt store on the right inboard pylon is a portable starter unit. (Nick Williams Collection)

By fall 1956 VF-74 added a red rudder with six stars and a red lightning bolt to the markings. The tail code was changed to AF. This Skyray is operating without drop tanks. (Nick Williams Collection)

A VF-74 Skyray on the flight line at the Naval Air Weapons Meet at NAAS El Centro, California. (Nick Williams Collection)

During a Mediterranean cruise aboard the USS **Roosevelt** ***(CVA-42) in 1957 VF-74 employed a tail code of AL and repainted the lightning bolt yellow with a yellow tail band.*** (Bill Davis via Nick Williams Collection)

(signifying Carrier Air Group Six). In the fall of 1956 the markings became more flamboyant—the rudder surfaces were painted red with six white stars, a red lightning bolt was painted across the forward fuselage, and 100-series nose numbers were applied.

By April 1957, the squadron was at full strength and went aboard the USS *Saratoga* (CVA-60) for carrier qualifications. The *Saratoga* was newly commissioned and was inspected by President Eisenhower during the carrier qualifications. While tied to the pier at Mayport, Florida, the new carrier launched two VF-74 Skyrays from the forward catapults in a dramatic display of modern naval air power for the visiting president.

In July 1957, VF-74 was deployed aboard the USS *Roosevelt* (CVA-42) for the first of three Mediterranean cruises. The squadron was now part of Carrier Air Group Seventeen. The tail code was changed to AL, the red markings were changed to yellow with black pinstripes, and 200-series nose numbers were applied. This was the first Skyray deployment with the Sixth Fleet, and the aircraft was still considered a day fighter.

VF-74 Skyrays participated in numerous exercises in the Atlantic and Mediterranean. LT Bill Davis flew for VF-74 during this cruise. He recalls that the Skyrays served as the Air Group's High Altitude Combat Air Patrol, "During these CAP exercises, we would fly on a radial out to 100 miles from the ship in 10- or 20-mile racetrack patterns, in flights of two at 35,000 feet. One was outbound all the time so that we could effectively extend the ship's radar coverage. Once during exercises over Crete, I was sent out to investigate a bogey. He was a high flyer, a British Vulcan bomber. I was on a trail with a couple of FJs. We had a chain behind him, so I stoked it into burner, which got me past the FJs. I got under the bogey enough to do a pitch-up to his altitude. The air is pretty thin up there and the sudden rotation was enough to disrupt the airflow, and I got compressor stalls. I eased out to let the air go straight in and she motored on up to the Vulcan, which was cruising at about 54,000 feet."[9]

Upon returning to port in February 1958 the squadron was based at NAS Oceana and was reattached to Carrier Air Group 6. A year later, VF-74 deployed aboard the USS *Intrepid* (CVA-11) for its second Mediterranean cruise, which included exercises with NATO, Italian, British, and Libyan forces. For this cruise the tail code was changed back to AF, the red markings were reapplied, 100-series nose numbers were used, and the name *USS Intrepid* was painted in red block letters on the rear of the fuselage.

The squadron returned to port in August 1959, and remained at NAS Oceana until August 1960, when they deployed for another Mediterranean cruise aboard the *Intrepid*. During this cruise, VF-74 pilots accomplished the first Skyray aerial refueling with a system that was devised by the squadron maintenance officer.

A retrofit of an A4D Skyhawk refueling probe to the nose of an F4D Skyray drop tank was the elegantly simple solution. The ComNavAirLant approved the modification, which was accomplished by NAS O&R

A close-up of the yellow lightning bolt worn by VF-74 Skyrays during the Mediterranean cruise aboard the USS* Roosevelt. *The squadron insignia can be seen on the rear portion of the canopy. (Nick Williams Collection)

The radome has been removed from this VF-74 Skyray to service the APQ-50 radar. (Bill Davis via Nick Williams Collection)

A VF-74 Skyray at the pier about to be hoisted aboard the USS* Intrepid *(CVA-11) for a Mediterranean cruise in 1959. The tail code has been changed back to AF and the red markings have been restored. (Nick Williams Collection)

***A VF-74 Skyray about to trap on the* Intrepid.** (Nick Williams Collection)

This VF-162 Skyray carries a refueling probe on its left drop tank that was developed by VF-74. It was an elegant solution to the Skyray's range problem. (U.S. Navy via Nick Williams Collection)

Jacksonville. Only two test flights of the refueling system could be conducted before the cruise, but the system worked flawlessly and within two months of deployment all VF-74 pilots were qualified in aerial refueling.

The system was subsequently approved for use in the Fleet and was adopted by several other Skyray squadrons. VF-74 returned from their final Mediterranean cruise in February 1961, and began transitioning to the new F4H-1 Phantom II in July. VF-74 operated F4D-1 Skyrays for more than six years—longer than any other Navy carrier-based unit.[10]

VF-13

VF-13 *Night Cappers*, based at NAS Cecil Field, Florida, transitioned to Skyrays in January 1959, much later than most squadrons. It was the last squadron to make an operational carrier deployment with the F4D. In April 1959 the squadron went aboard the USS *Randolph* (CVA-15) for carrier qualifications as part of Carrier Air Group Ten. VF-13's first Skyray deployment was from August 1959 until February 1960 aboard the USS *Essex* (CVA-9). This was the first Skyray deployment aboard a carrier with outdated hydraulic catapults. During this Mediterranean cruise the VF-13 Skyrays carried the tail code AK. The squadron markings were rather elaborate. The elevons were painted with red and white stripes with a black star centered in each stripe, and the lower half of the rudder was red and black separated by a diagonal black bar.

VF-74 Skyrays parked on the deck of the USS* Intrepid *(CVA-11) during a 1959 Mediterranean cruise. (Tailhook Association via CDR Doug Siegfried)

Upon returning from the *Essex* Mediterranean cruise, the squadron reported to the USS *Shangri-La* (CVA-38) for exercises off the coast of Cuba, which lasted until September 1960. In February 1961, VF-13 adopted the in-flight refuelable drop tanks that had been developed by VF-74 and set out on another Mediterranean cruise aboard the *Shangri-La*. But this cruise was shortened by the Dominican Crisis, which resulted in the carrier's return to the Caribbean.

In February 1962, VF-13 deployed aboard the *Shangri-La* for the squadron's third Mediterranean cruise. By now the Fleet F4Ds were growing old; this would be the last operational carrier cruise for the Skyray. LT Rod Creagh was among the VF-13 pilots on that final Skyray deployment and he recalls how worn the Skyray airframes had become, "On 10 July 1962 Carrier Air Wing Ten was conducting day flight operations off the USS *Shangri-La* (CVA-38) in the Mediterranean Sea, south of the coast of France. I was part of a two-plane section with LTJG Dick Borgula, flying the F4D-1 BuNo 139050. After an uneventful flight, we returned to the carrier, and I made my approach for landing. The seas were relatively calm and the deck was stable.

"Upon contacting the arresting wire, I felt the initial resistance, followed by release—my thoughts were that the hook point had pulled off, and that I would have to bingo [divert] to Cagliari—and miss going to Cannes for some much awaited shore leave.

"At that point the Air Boss called on the VHF saying '105, you appear to be on fire.' This was somewhat disconcerting for my wingman who was flying 105, since my aircraft side number was 106. I then saw the fuel gauge dropping *rapidly* and announced my decision to climb and eject. It was necessary to lower the seat in the aircraft and extend the pedals to assume a position in the cockpit, which would lessen the chance for cervical or lower back injury from ejection. This was easily done while still climbing using toggle switches on the left console.

VF-13 acquired their Skyrays in the winter of 1959. They used elaborate markings during their first cruise aboard the USS* Essex *(CVA-9). The lower rudder was painted red and black, and the elevons carried seven red and six white stripes. The squadron tail code was AK. This is a VF-13 Skyray landing aboard the* Essex *in 1960. (U.S. Navy via Nick Williams Collection)

In the spring of 1960 VF-13 made a North Atlantic cruise aboard the USS* Shangri-La *(CVA-38) when this two-ship formation was photographed. (U.S. Navy via Nick Williams Collection)

A VF-13 Skyray in full afterburner prepares to launch from the USS* Shangri-La *(CVA-38) during a 1961 Mediterranean cruise. (U.S. Navy via Nick Williams Collection)

"At approximately 800 to 1,000 feet, I ejected and the Martin-Baker seat performed beautifully (thankfully). On the way down, I saw the airplane gliding above me, turning downwind where it hit the water at roughly the 180 position. The helicopter was already airborne, as they always are for daylight flight operations, and picked me up without incident, using the 'horse collar' sling. It was nice to be on board! A checkup by our Flight Surgeon [Waller Tabb, M.D.] showed no spinal injuries and nothing more than a sore back. I did enjoy shore leave in Cannes, and 10 days later I was back in the flight schedule."[11]

Unfortunately, tail hook assembly failures were becoming common; VF-102 experienced one aboard the Forrestal approximately one year before LT Creagh's incident.

CDR Hjalmer E. Swanson, the Executive Officer of VF-13, flew the last operational carrier flight of an F4D-1 Skyray in August 1962. The squadron returned to NAS Cecil Field in September and exchanged their Skyrays for F3H-2 Demons.[12]

VF-51

Based at NAS Miramar, VF-51 was the shortest-lived Skyray unit, operating the F4D-1 for less than two years. The *Screaming Eagles* of VF-51 began to transition from Grumman F11F-1 Tigers to F4D-1 Skyrays in January 1959. Initially their aircraft wore subdued markings, which consisted of the tail code NF, with VF-51 at the base of the fin, a nose number on the fuselage, and a red fin cap. In mid 1959 the squadron pilots convinced their Commanding Officer that more colorful markings were needed for a squadron named the *Screaming Eagles.* Three stylized talons were painted in red on both sides of the fuselage and the drop tanks. However, COMNAVAIRPAC disapproved of the flamboyant markings and requested their removal.

A VF-13 Skyray taking a wave-off from the USS* Shangri-La *(CVA-38). (U.S. Navy via Nick Williams Collection)

VF-51 had a short association with the Skyray. In mid 1959 the squadron added a colorful red talon design to their F4Ds. (Tailhook Association via CDR Doug Siegfried)

A VF-51 Skyray taxies to a catapult aboard the USS* Ticonderoga *(CVA-14). (Tailhook Association via CDR Doug Siegfried)

VF-51 Skyrays on their only Skyray cruise aboard the USS* Ticonderoga *(CVA-14) in the spring of 1960. (U.S. Navy via Nick Williams Collection)

Ground personnel remove graffiti from a VF-51 Skyray that inadvertently landed at the wrong airfield. (Tailhook Association via CDR Doug Siegfried)

In drab markings once again, VF-51 boarded the USS *Ticonderoga* (CVA-14) in October 1959 for carrier qualifications. In March 1960 VF-51 deployed on a WestPac cruise aboard the *Ticonderoga* attached to Carrier Air Group Five and returned to port in October 1960. By November 1960 they ended their short venture with the Skyray, and began to convert to Vought F8U Crusaders.[13]

In the spring of 1961 VF-162 was assigned to Carrier Air Group Six. Its modex was changed to AF and it prepared to replace VF-74 aboard the USS Intrepid (CVA-11). This photograph was taken at NAS Cecil Field, Florida, prior to the deployment. (Tailhook Association via CDR Doug Siegfried)

VF-162

VF-162 was established at NAS Cecil Field, Florida, on 1 September 1960 along with four other squadrons and a parent organization, Carrier Air Group Sixteen. This air group and squadrons were formed to serve during a transitional period, and had funding to exist for only one year. VF-162 was the only F4D squadron in the group, and its Skyrays were provided by units that were transitioning to newer fighters. VF-162 was expected to provide fill-in all-weather fighter service for units that were temporarily occupied with training on the McDonnell F4H Phantom II. VF-162 built up quickly with the first Skyray arriving at Cecil Field on 10 October 1960. By late December the unit had a full complement of aircraft and nearly a full complement of pilots. An intense training period then began.

During the winter of 1960 through the spring of 1961 several detachments of VF-162 were deployed to NAS Key West, Florida, to stand alert during the Cuban Missile Crisis, replacing the Skyrays

VF-162 was formed to serve an interim all-weather fighter role in 1960. This Skyray, which carries the squadron's initial tail code AH, is one of the first received by VF-162. (Nick Williams Collection)

A VF-162 Skyray aboard the **Intrepid** ***in 1961. The object on the nose of the left drop tank is an A4 refueling probe. It was used on all VF-162 Skyrays, and proved to be an effective means of giving the Skyray refueling capability.*** (Nick Williams Collection)

A VF-162 Skyray taxies on the deck of the USS **Intrepid** ***(CVA-11) in the spring of 1961. VF-162 provided an all-weather fighter squadron replacement for VF-74, which was transitioning to F-4 Phantom IIs.*** (Tailhook Association via CDR Doug Siegfried)

A three-ship formation of VF-162 Skyrays from the USS* Intrepid *(CVA-11) during its Mediterranean cruise on 17 August 1961. (U.S. Navy via Nick Williams Collection)

of VF-101. At this time VF-162 Skyrays wore an elaborate color scheme that featured a black rudder and elevon randomly sprinkled with gold stars. The tail carried an AH modex and a 200-series nose number was applied near the radome. The squadron's *Mighty Hunter* insignia appeared on the rear canopy. The drop tanks also carried the black field with gold stars, and the refueling probe devised by VF-74 was employed on VF-162 Skyrays.

In spring 1961, VF-162 transferred to Carrier Air Group Six, changed its modex to AF, and replaced VF-74 aboard the USS *Intrepid* (CVA-11). Between 15 April and 1 August 1961 the squadron made four deployments. The first and second deployments were carrier qualifications off the coast of Norfolk, Virginia. The third was in response to an international crisis that resulted from the assassination of the Dominican Republic dictator, Trujillo. The fourth was an Atlantic Fleet exercise conducted off the coast of New York.

VF-162's final deployment was a Mediterranean cruise aboard the USS *Intrepid* (CVA-11) from 3 August 1961 to 28 February 1962. After returning from the cruise in April 1962, the squadron began to prepare their Skyrays for delivery to the NAS Jacksonville Overhaul and Repair Depot, marking the end of their mission.[14]

VF-23

VF-23 based at Moffett Field, California, acquired its first F4D-1 Skyray on 11 December 1956. The squadron had been equipped with McDonnell F2H-3 Banshees and was attached to Air Task Group Four Western Pacific. After 10 months of intensive training with the F4D, the squadron went aboard the USS *Hornet* (CVA-12) for carrier qualifications on 20 September 1957. VF-23 markings consisted of a bright yellow tail with the ATG-4 modex ND in black. Two horizontal black bands ran across the top and bottom of the yellow field, and the full aircraft BuNo was painted at the base of the yellow field. A 200-series nose number was painted aft of the radome, and the national insignia was painted farther aft than other Skyray units.

VF-23 received its first Skyrays in December 1956 at Moffett Field. The squadron used a tail code of ND, which was placed on a large yellow field. Unusual features of VF-23's markings were the use of the entire BuNo on the tail, and large nose number with the rearward placement of the National Insignia causing a portion of it to be painted on the inlet ramp. (Top: Nick Williams Collection; middle and bottom: Bill Larkins)

In September 1957 VF-23 flew carrier qualifications aboard the USS* Hornet *(CVA-12). During these operations the Skyrays operated with no drop tanks. (Mark Nankivil Collection)

LT Arland Nankivil in the cockpit of a VF-23 Skyray during a training exercise. LT Nankivil would be selected to fly with the elite VF(AW)-3 squadron after his tour with VF-23. (Mark Nankivil Collection)

A VF-23 Skyray is refueled at NFLP Crow during training exercises in 1957. (Mark Nankivil Collection)

This VF-23 Skyray was photographed at NAS Miramar in November 1957. The squadron flew to Miramar for Sidewinder missile practice. (Mark Nankivil Collection)

The squadron completed carrier qualifications and returned to Moffett Field on 11 October 1957. On 29 October, VF-23 was detached from ATG-4 and assigned to Carrier Air Group Fifteen. On 3 November 1957 they flew to NAS Miramar, California, for Sidewinder missile training. During December 1957 and January 1958 they flew additional carrier qualifications from the USS *Hancock* (CVA-19). From 26 to 31 January the squadron flew to NAAS Fallon, Nevada, for gunnery practice.

On 15 February 1958, VF-23 deployed aboard the USS *Hancock* for its first—and last—Skyray cruise. The aircraft markings were changed to a tail code of NL in slanted black letters, with two yellow rays replacing the earlier solid yellow field. The last four digits of the BuNo were located at the base of the tail, and the drop tanks received a black nose and yellow rays.

During the WestPac cruise VF-23 participated in a weapons demonstration for Asian military leaders from 20 to 23 May, Operation Knockout from 9 to 17 June, and provided Combat Air Patrol against simulated air strikes from the USS *Shangri-La* (CVA-38) in July. VF-23 also participated in Blue Sky, a joint U.S./Nationalist Chinese exercise in August 1958.

The squadron returned to Moffett Field on 3 October 1958 where they received a commendation from the Commander of Carrier Division One, RADM Paul H. Ramsey, for outstanding performance during the cruise. VF-23 began to transition into McDonnell F3H-2 Demons, and by January 1959 they ceased Skyray operations.[15]

VF-101

Based at Cecil Field, Florida, VF-101 adopted the nickname of a famous World War II squadron, *The Grim Reapers*. VF-101 was commissioned in 1952 as an F2H-3 Banshee unit, and received their first Skyray in August 1956. They operated that type until 1962, making it one of the longest tenured Skyray units in the Fleet. The squadron's initial F4Ds were not radar equipped, and were used for familiarization training. When the first radar equipped Skyrays arrived in early 1957 VF-101 began an intensive program in all-weather tactics. In May 1957 they became the first Skyray squadron in the Atlantic Fleet to complete day carrier qualifications aboard the USS *Roosevelt* (CVA-42). They completed night qualifications in July 1957 aboard the USS *Saratoga* (CVA-60).

Early VF-101 Skyrays carried only a tail code of J and a 100-series nose number. In 1957 the squadron applied a red band with four white stars to the tail and similar markings to the drop tanks. On 3 September 1957 they changed the tail code to AP and deployed aboard the *Saratoga* for a joint NATO exercise, Operation Strikeback. During this operation VF-101 participated in cross-deck exercises with the British carrier, HMS *Ark Royal*.

On one occasion a VF-101 Skyray lost its catapult holdback fitting when launched from the *Saratoga*. The pilot, LT Max K. Morris, landed aboard the *Ark Royal* and made a free deck launch from that ship using only 580 feet of deck length. The takeoff was made using afterburner while carrying a light fuel load. VF-101 returned to Cecil Field on 22 October 1957 and redeployed for a short VIP cruise aboard the USS *Forrestal* (CVA-59) two weeks later.

In April 1958 the squadron relocated to NAS Key West, Florida, and merged with the Fleet All Weather Training Unit, Atlantic (FAWTULANT). It was assigned the mission of training replacement pilots for duty with Fleet units using both Demons and Skyrays. At this point VF-101 had grown quite large with an inventory of 36 aircraft (11 F3D Skyknights, 11 F3H Demons, and 14 F4D-1 Skyrays). By the spring of 1959 the inventory would expand to 64 aircraft. In January 1959 VF-101 pilots participated in a series of Big Blast exercises, intercepting B-47 penetrations in their warning area.

VF-101 was instrumental in developing many training aids used in Skyray all-weather interceptor practice: the Dart target one-way tow reel, the Del Mar target tow rig, a four-shot reusable rocket pod, a radar optic scorer

In 1960, VF-101 was attached to replacement Carrier Air Group Four and wore these markings. (U.S. Navy via Nick Williams Collection)

recording camera installation, and a Sidewinder adaptor for the Aero 20-A bomb rack. The squadron supplied extensive target towing services for the Fleet including competitions such as the Naval Air Weapons Meet and Operation Top Gun. VF-101 continued to use their Skyrays training all-weather replacement pilots throughout 1960–1961 and the tail code was changed to AD signifying their attachment to Replacement Carrier Air Group Four. In 1962, The Grim Reapers transitioned to the McDonnell F4H Phantom II.[16]

VF-141

NAS Miramar-based VF-141 *Iron Angels* received their first Skyray on 26 June 1956. The squadron flew carrier qualifications aboard the USS *Essex*, *Lexington*, and *Bon Homme Richard*. While aboard the *Lexington* a VF-141 pilot made the first successful underwater ejection from a Skyray. After making a successful arrested landing LT Sothan taxied forward and applied his brakes. Not knowing that his right brake was inoperative he taxied off the port side. The F4D hit the water, inverted, and sank to a depth of approximately 70 feet. He initiated the ejection sequence and surfaced 1,000 feet aft of the ship where he was recovered uninjured.

Initially, VF-141 Skyrays carried a Carrier Air Group Five modex of S on a large yellow pennant with 200-series nose numbers. On 12 July 1957 VF-141 embarked on a WestPac cruise aboard the USS *Bon Homme Richard* (CVA-31) and the tail code was changed to NF on some aircraft. The squadron returned to NAS Miramar on 9 December 1957 and most of its pilots received orders to new duty stations. There, former VF-141 pilots remained to indoctrinate new pilots reporting to the squadron. The squadron reformed efficiently, and when called upon for joint exercises in May 1958, 100 percent of the assigned missions were completed. In August 1958 the *Iron Angels* were assigned to Carrier Air Group Fourteen, and the tail markings were changed once again. The tail code NK was painted in slanted white letters over an orange ray with black borders.

From 21 to 27 September 1958 VF-141 went aboard the USS *Ranger* (CVA-61) for day carrier qualifications. On 11 to 17 October the squadron returned to the USS *Ranger* for day and night qualifications. On 28 December 1958 VF-141 deployed aboard the *Ranger* for their second WestPac cruise. The squadron participated in operational readiness training at sea and underwent an Operational Readiness Inspection from 9 to 13 February 1959. The unit spent 16 to 28 March at NAS Atsugi, Japan, and returned to the *Ranger* for operations at sea. On 25 to 30 June 1959 VF-141 Skyrays served as Carrier Air Group Fourteen's all-weather squadron during the Granite Creek exercise. On 27 July 1959 the Iron Angels returned to NAS Miramar and relinquished their F4D-1 Skyrays for F3H-2 Demons.[17]

A VF-141 Skyray is brought to the flight deck aboard the USS* Bon Homme Richard *(CVA-31) during a WestPac cruise in 1957. (Boeing via Nick Williams Collection)

VMF(AW)-114

The *Death Dealers* of VMF(AW)-114 were commissioned at MCAS El Toro, California, in 1943 as a day fighter F4U Corsair squadron. They were transferred to MCAS Cherry Point, North Carolina, in 1946 with a change of mission to a night fighter unit. On 1 May 1957 they were redesignated an all-weather squadron as they converted from F9F-8 Cougars to F4D-1 Skyrays. In June 1957 VMF(AW)-114 began an extensive program of shore training and carrier qualifications with their new Skyrays. Initially, the squadron applied a tail code of EK to their aircraft to signify their attachment to the Second

A VF-141 Skyray about to catch the number-three wire of the USS* Bon Homme Richard *(CVA-31) in December 1957. (U.S. Navy via Nick Williams Collection)

Marine Aircraft Wing, Fleet Marine Force, Atlantic. On 13 February 1959 the squadron deployed aboard the USS *Roosevelt* (CVA-42) for a Mediterranean cruise. The tail code was changed to AB for this deployment indicating attachment to Carrier Air Group One. During the cruise VMF(AW)-114 lost a pilot and aircraft, but nevertheless received the Marine Corps Commandant's annual Aviation Efficiency Trophy, awarded for flying efficiency and safety. During this time VMF(AW)-114 became the second Marine Skyray squadron to become day and night carrier qualified.

Returning to MCAS Cherry Point after the cruise, the aircraft markings were changed again. This time the squadron's EK tail code was embellished with a pattern of high-visibility Fluorescent Red-Orange on the tail and spine, similar to the VMF(AW)-115 scheme. The squadron's *Death Dealer* insignia was applied to the rear of the canopy.

In February 1961 the squadron was assigned to Provisional Marine Aircraft Group Ten at NAS Roosevelt Roads, Puerto Rico, in support of the 24th Marine Expeditionary Unit, which conducted amphibious operations on Vieques Island, Puerto Rico. During this assignment a VMF(AW)-114 Skyray became the first jet aircraft to land on the island's Strategic Expeditionary Landing Field (SELF) on 6 February 1961.

The squadron returned to MCAS Cherry Point for the remainder of 1961, but in the summer of 1962 they deployed to NAS Atsugi, Japan, where they were attached to Marine Air Group Eleven, 1st Marine Air Wing, Fleet Marine Force, Pacific. Again the aircraft markings were changed to a slanted EK tail code with a playing card motif of spade, club, heart, and diamond painted on the rudder. The drop tanks received a red lightning bolt. The squadron remained based at Atsugi through 1963 with short deployments to NAF Naha,

VMF(AW)-114 F4Ds prepare to be launched on the* Roosevelt *in 1959. (Tailhook Association via CDR Doug Siegfried)

In 1960 VMF(AW)-114 changed their markings to incorporate a Fluorescent Red-Orange design. The tail code reverted to the original EK. (U.S. Marine Corps via Nick Williams Collection)

Okinawa, and Ping Tung, Taiwan. On 1 July 1963 VMF(AW)-114 was disbanded at Atsugi, but there are photographs of VMF(AW)-114 Skyrays in new markings at MCAS Cherry Point as late as October 1963. This suggests that the squadron reformed after the Atsugi deployment; however, there is no official account.[18]

VMF (AW)-531

VMF(AW)-531 was commissioned on 16 November 1942 as the Marine Corps' first night fighter squadron. Nicknamed the *Grey Ghosts,* the squadron received its first Skyray at MCAS Cherry Point, North Carolina, on 22 February 1958. During the next year the squadron trained with the F4D in preparation for an extended deployment to the Far East. In March 1959 VMF(AW)-531 began its relocation to NAS Atsugi, Japan, and in April 1959 the squadron took possession of the F4Ds left behind by VMF(AW)-115. In September 1959 the squadron underwent carrier qualifications aboard the USS *Lexington* (CVA-16), and again in December aboard the USS *Midway* (CVA-41). During the carrier qualifications the aircraft wore a tail code of EC and the unit designation on the spine.

In January 1960 the squadron deployed to NAS Cubi Point, Philippines, and in March they operated briefly from Ping Tung, Taiwan, in an exercise named Operation Blue Star. During this exercise the VMF(AW)-531 Skyrays operated off of the Strategic Expeditionary Landing Field where they made Morest landings and steep afterburner takeoffs from the temporary steel matting runway.

In late May 1960 the squadron returned to NAS Atsugi before being cadred in June. The squadron reformed at MCAS Cherry Point, and the process of retraining was begun. More carrier qualifications were flown aboard the USS *Intrepid* (CVA-11) with newly

In the summer of 1962 the squadron carried these markings during a deployment to NAS Atsugi, Japan. (Fred C. Dickey, Jr., via Nick Williams Collection)

A VMF(AW)-531 Skyray during carrier qualifications aboard the USS* Lexington *(CVA-16) in September 1959. (Nick Williams Collection)

acquired Skyrays (the former aircraft remained at Atsugi). The markings were similar to the earlier scheme; however, the EC tail code was now slanted.

In July 1961 VMF(AW)-531 deployed to NAS Atsugi for a second time. Their markings reverted to the upright block letter EC tail code, but the drop tanks carried a DayGlo Orange nose. VMF(AW)-531 completed this second Japan deployment in July 1962 and returned to MCAS Cherry Point to transition into the McDonnell F-4B Phantom II.[19]

VMF(AW)-542

VMF(AW)-542 was originally organized as a night-fighter squadron in March 1944. Based at MCAS Cherry Point, North Carolina, it was designated VMF(N)-542. In July 1947 the squadron relocated to MCAS El Toro, California, and was redesignated an all-weather squadron. After flying Grumman F7F-3N Tigercats for several years, VMF(AW)-542 transitioned to the F3D-2 Skyknight in August 1957 and the F4D-1 Skyray in May 1958. The squadron's Skyray pilots trained at El Toro for the next year and underwent carrier qualifications aboard the USS *Bon Homme Richard* (CVA-31) in July 1959. At

VMF(AW)-531 Skyrays operating from a Strategic Expeditionary Landing Field (SELF) during Operation Blue Star on 27 March 1960. (U.S. Marine Corps via Nick Williams Collection)

VMF(AW)-531 Skyrays operating from NAS Atsugi, Japan, in January 1962. (U.S. Marine Corps via Nick Williams Collection)

this time the Skyrays wore a tail code of WH painted in small connected letters on the rudder.

On 20 August 1959 the squadron deployed to the Far East replacing VMF(AW)-314 at NAS Atsugi. During this deployment the unit traveled to NAS Cubi Point, Philippines, from January to May 1960. While at Cubi Point they participated in a joint American–Nationalist Chinese amphibious exercise, Operation Blue Star.

In November 1960 VMF(AW)-542 was rotated back to MCAS El Toro where they performed various training missions with their Skyrays. They made temporary deployments to Yuma, Arizona, and to Point Mugu and El Centro, California. The squadron tail code remained WH, but it was now rendered in large block letters that covered much of the tail. The drop-tank noses and wingtips were painted dark blue. Later in 1962 small white stars were added to the drop-tank markings.

On 30 October 1962 the squadron was sent on their second Far East deployment with their Skyrays (now designated F-6A under the new joint service nomenclature). In 1963 the squadron visited Okinawa and Taiwan before returning to MCAS El Toro and transitioning into their new McDonnell F-4B Phantom IIs.[20]

VMF(AW)-513

Nicknamed the Flying Nightmares and based at MCAS El Toro, California, VMF(AW)-513 received their first Skyrays in July 1958. At that time the squadron was attached to the Third Marine Aircraft Wing, Marine Air Group Fifteen. From 1959 to 1961 VMF(AW)-513 operated their Skyrays on short deployments to various training fields in the southwest. They flew to MCAAS Yuma, Arizona, for gunnery and rocket-firing practice. While at Yuma they also participated in the annual Operation Top Gun competition. The squadron markings during this period consisted of a block-lettered tail code WF with VMF(AW)-513 painted on the dorsal spine, and single- or double-digit nose numbers.

During this period Gary McClellan, a former American Airlines captain, currently a test pilot for Boeing, reported to VMF(AW)-513 to fly Skyrays. He

VMF(AW)-542 received its first Skyray at MCAS El Toro, California, on 12 May 1958. The early markings consisted of a WH tail code painted in small slanted letters on the rudder of the aircraft. (U.S. Marine Corps via Nick Williams Collection)

This VMF(AW)-542 Skyray received special markings, which included a red arrow with white stars on the tail and red elevon tips. (U.S. Marine Corps via Nick Williams Collection)

A VMF(AW)-542 on final approach to NAS Atsugi, Japan, during the squadron's second Far East deployment in 1962. (Tailhook Association via CDR Doug Siegfried)

In November 1960, after returning from a Far East deployment, VMF(AW)-542 Skyrays received new markings, which consisted of a larger block-lettered WH tail code, with a dark blue fin cap and drop-tank nose. (U.S. Marine Corps via Nick Williams Collection)

In late 1962 the squadron markings were enhanced again to include a spiral band of stars painted on the blue drop-tank nose. (Doug Olson via Nick Williams Collection)

VMF(AW)-513 acquired its first Skyray on 26 July 1958. The early markings consisted of a block-lettered tail code WF. Some aircraft carried red fin caps and wingtips, while others were black. (U.S. Marine Corps via Nick Williams Collection)

On a Far East deployment to NAS Atsugi, Japan, in 1961-1962, VMF(AW)-513 enhanced their markings to include dark blue fin caps, wingtips, and drop-tank noses with three gold stars. (bottom: M. D. Stride via Nick Williams Collection; top: T. Matsuzaki via Nick Williams Collection)

recalled the training that molded him into a proficient Ford pilot, "I was a young Marine 2dLt straight out of flight school with a total of 331 hours, 27 in the Grumman F-11F Tiger, so I at least knew what a high-performance fighter looked like. The Skyray checkout consisted of an in-depth familiarization with the F4D operating manual and 15 hours in the cockpit procedures trainer. The next adventure was my first solo [single-seat] flight in the Skyray and what an impressive ride it was.

"On takeoff roll I applied military power, started rolling, engaged afterburner, the acceleration was unbelievable, 0 to 160 knots in 10 seconds. The secret to a successful first flight was to get the gear up before exceeding the gear limit speed. The flight objective was general familiarization with the low-speed and high-speed characteristics and maneuverability of the Skyray. Low-speed flight was at a high angle-of-attack with a relatively high power setting. This configuration provided stability, airspeed, and sink rate control for carrier landings. The forward visibility was excellent, even at the nose-high attitude. The high-speed and high-altitude maneuverability were superior to most aircraft.

"I flew the Skyray from July 1960 until November 1962, training the first year with VMF(AW)-513 at MCAS El Toro, California. The training included air-to-air intercepts, air-to-air combat maneuvering, formation flying, air-to-ground weapons delivery, cross-country flying, and day and night carrier landings. The Skyray armament consisted of an APQ-50 radar system capable of acquiring a target, tracking it automatically, and supplying tracking information to the pilot, which enabled him to carry out an effective lead collision or lead pursuit intercept. The fire control system provided accurate control of four 20mm guns, four external rocket packs, or four Sidewinders. The 513 pilots became very proficient in all modes of weapons delivery.

"The most exciting and challenging aspect of flying a Navy or Marine pilot encounters are day and night carrier landings. Night carrier landing is unarguably the most intense challenge. Coming aboard the boat in the Skyray was no exception.

"The rapid deceleration during a trap, from 138 knots to a dead stop in the length of an angle carrier deck is something one has to experience, it can't be described. The same goes for a catapult shot, 0 to 160 knots in a few seconds. As I recall, you are positioned on the catapult, run the engine up to full military power, select afterburner, salute the Catapult Officer, and become instantly airborne at 160 knots.

"In November 1961, VMF(AW)-513 deployed to the Far East. We spent a year experiencing the best flying in the world. Our squadron was the finest Skyray squadron in the Marine Corps. Even today we have a squadron reunion in conjunction with the Marine Corps Aviation Association annual meeting. Those years hold the greatest memories of my aviation career."[21]

As Gary McClellan recalled, VMF(AW)-513 deployed to NAS Atsugi, Japan, on 16 October 1961. The squadron was assigned to the 1st Marine Air Wing, Marine Air Group Eleven. While at Atsugi the squadron adopted more colorful markings, which featured dark blue tail and elevon tips with three gold stars. A similar design was painted on the drop tanks.

In November 1962, VMF(AW)-513 returned to the States and turned in their Skyrays for F-4B Phantom IIs.[22]

A young 2dLt Gary McClellan transitioned to the F4D after only 27 hours in high performance jets. Currently a test pilot for Boeing, he regards his Skyray experience in VMF(AW)-513 as the finest flying of his career. Aircraft shown here is from VMF(AW)-542. (Nick Williams Collection)

Chapter Seven

Utility, Development, Training and Reserve Units

Five VU-3 Skyrays in formation flight in January 1964. The variations in markings among the aircraft can be seen. Fin cap colors and drop-tank markings differ, also some carry squadron logos on their canopy. Some have their BuNo in large block letters painted on the fin, while others carry small letters on the rudder. (U.S. Navy via Nick Williams Collection)

Located on both the East Coast and the West Coast to augment front-line operational squadrons were the Utility, Development, Training, and Reserve units that served a variety of important functions in support of the Skyrays in the fleet. Adorned with unique and colorful markings, their F4Ds became the last Skyrays to fly.

VU-3

Utility Squadron Three was commissioned in December 1948 at NAS Santa Ana, California, to fulfill the mission of providing radio controlled drone aircraft for Fleet target practice. After relocating to NAS Miramar in 1950, NAAS Ream Field in 1951, and NAAS Brown Field in 1956, the squadron settled at NAS North Island, San Diego, in 1961. They began operating Skyrays in 1962 when the F4D designation was changed to F-6A under the Department of Defense joint service redesignation order. As drone controllers, the VU-3 Skyrays were technically listed as DF-6As, but that designation was never painted on the aircraft. High-visibility markings were used to protect the VU-3 Skyrays during target exercises. Tails, outer wing panels, drop tanks, and forward fuselages were painted Fluorescent Red-Orange. However, the squadron markings were not standardized, and patterns varied on each Skyray.

In addition to providing drone controller service for target practice, VU-3 Skyrays were used to provide air intercept controller training for personnel on board Fleet destroyers, cruisers, carriers, and personnel in training at the Fleet Anti-Air Warfare Training Center.

Skyrays were being retired from frontline squadrons in 1962 and they were put to use in utility squadrons such as VU-3. They retained the standard Gull Gray and white tactical finish, but they received DayGlo Red-Orange wingtips, tails, and noses. Some VU-3 Skyrays carried DayGlo markings on their drop tanks. (U.S. Navy via Nick Williams Collection)

Douglas attempted to sell Skyrays to the Indian government in the early 1960s. Two VU-3 Skyrays, BuNos 139122 and 139153, were used as demonstrator aircraft. Douglas was unsuccessful in obtaining foreign orders. (Boeing via Nick Williams Collection)

In May 1964, VU-3 flew two F-6A Skyrays to the Douglas plant in Long Beach, California, to serve as demonstrator aircraft for the Indian government. Douglas attempted to secure foreign orders for the aging fighter, but none were forthcoming.

VU-3 was the last operational squadron to fly the Skyray. A few F-6As would continue to fly test, developmental, and training missions, but it was officially retired from operational use on 16 October 1964. An honorary retirement ceremony was held after its last operational mission, then VU-3 squadron pilots ferried their Skyrays on a final flight to NAF Litchfield Park, Arizona, for disposal.[1]

VX-3

VX-3 was the East Coast development squadron charged with the mission of evaluating new and already existing naval aircraft, airborne equipment and methods for the most effective tactical employment, and to recommend training procedures and countermeasures for these aircraft and methods. The squadron was formed on 20 November 1948 and placed under the control of Commander Operational Development Force. It was based at NAS Atlantic City, New Jersey, and was staffed by some of the Navy's most experienced aviators who evaluated virtually every carrier aircraft in the Fleet.

VX-3 received its first Skyray on 8 January 1957 and began a thorough operational evaluation of the aircraft, weapons, and equipment. In April 1957, VX-3 flew its Skyray aboard the USS *Roosevelt* (CVA-42) for carrier qualifications and advanced Tactical Aircraft Control and Navigation (TACAN) work. In June 1957 the project was moved to the USS *Saratoga* (CVA-60) for all-weather TACAN evaluations. By the fall of 1957 VX-3 was operating seven Skyrays in various evaluation projects. In 1958, VX-3 pilots made use of the USS *Forrestal* (CVA-59) and the USS *Independence* (CVA-62) while operating out of NAS Oceana, Virginia.

VX-3, based at NAS Atlantic City, New Jersey, received its first Skyray on 8 January 1957. The markings were very plain consisting of an XC tail code, which was later changed to JC in 1958. (U.S. Navy via Peter Bowers Collection)

A VX-3 Skyray being prepared for launch from the USS* Forrestal *(CVA-59) in April 1958 during a TACAN evaluation project. The tail code of VX-3 had been changed to JC. (U.S. Navy via Robert Lawson)

VX-3 Skyrays carried a tail code of XC, and VX-3 painted on the spine with a nose number located aft of the radome. The squadron was scheduled to relocate to NAS Roosevelt Roads, Puerto Rico, in February 1960 but it was decommissioned on 1 March 1960 and its Skyray inventory was transferred to other Fleet squadrons.[2]

VX-4

Air Development Squadron Four was formed on 15 September 1952 at the Naval Missile Center, Point Mugu, California. Its mission was to develop and conduct evaluations of air-launched guided missiles, and to develop all-weather intercept tactics for the best use of those missiles. Initially, VX-4 was involved in Sparrow I launches from Vought F7U-3 Cutlasses and Douglas F3D Skyknights. With the advent of the Bullpup and Sparrow III missiles, the squadron acquired FJ Furies, A4D Skyhawks, F3H Demons, F8U Crusaders, and F4D-1 Skyrays for advanced missile test programs.

VX-4, based at Point Mugu, California, used Skyrays for advanced missile test programs. VX-4 used very sparse markings, which consisted of an XF tail code. (U.S. Navy via Nick Williams Collection)

California. Its mission was to provide aerial gunnery training for Fleet pilots. The unit operated F8U-1s, FJ-4Bs, A4D-2s, and F4D-1s as target tugs and gunnery aircraft. The FAGUPAC postgraduate course consisted of five weeks of classroom and flying instruction on all-weather air-to-air and air-to-ground weapons delivery and tactics. Sixty hours were spent in class and 40 hours were spent in flight.

FAGUPAC received its first Skyray in January 1958, and applied the squadron's distinctive scheme of a red and white checkerboard tail with the tail code TR superimposed over it. Two black bands encircled the aircraft nose with a red lightning bolt arcing through them. The nose of the drop tanks carried a similar pattern. At some later date the size of the checkerboard field was reduced to a small band running across the upper portion of the tail. In summer of 1959 the squadron was relocated to MCAAS Yuma, Arizona, and by the end of that year FAGUPAC relinquished their Skyrays.[4]

Reserve Squadrons and Shore Facilities

Another old saying is, "Military aircraft never die, they just get sent to the Reserves." While generally true, the move of F4D Skyrays from operational units to Naval Reserve squadrons actually breathed new life into the career of this unique Navy jet fighter.

VF-881, 882, VMF-113, 215

In 1962, as Skyrays were being withdrawn from frontline service

In 1962 the reserve squadrons at NAS Olathe, Kansas, transitioned into F4D-1 Skyrays. The Skyrays represented a large increase in performance over the non-afterburning jets that had been used in the reserve squadrons, but the pilots adapted quickly. (U.S. Marine Corps via Nick Williams Collection)

they were released to reserve fighter units at NAS Olathe, Kansas. Approximately 18 Skyrays were accepted by VF-881 and VF-882 of the Navy's Reserve Air Wing 88, and VMF-113 and VMF-215 of the Marine Air Reserve Training Detachment Olathe. The airplanes received Fluorescent Red-Orange markings over their standard Gull Gray and white paint schemes. The tail code 7K was painted on the upper fin, and both *Navy* and *Marine* was painted on the lower fin and fuselage. However, some Olathe Skyrays remained gray and white, never receiving the DayGlo trim.

By now the F4D designation had been officially changed to F-6A, and the aging Skyrays proved to be a handful for the reserve units. Most reservists' jet experience was in slower non-afterburning F9F-8 Cougars, and the Skyray was a quantum leap forward in performance. Spare parts and technical assistance were limited, so systems such as the autopilot were often down. Several Skyrays developed hairline cracks in their afterburner fuel lines resulting in airborne fires. At one point, all Skyrays were grounded to inspect and replace J57 afterburner fuel lines. The yaw damper system became another source of failure and caused the only reserve Skyray fatality. The yaw damper on LT Donald K. Cauble's aircraft unexpectedly deflected hard over during takeoff while the Skyray was in full afterburner resulting in an inverted impact.

In spite of these difficulties, all four Olathe reserve squadrons become highly proficient with the F-6A Skyray and won numerous commendations from their commands. In 1964, the Navy units VF-881 and VF-882 transitioned to A-4B Skyhawks while the Marine units, VMF-113 and VMF-215, converted to F-8A Crusaders in 1965.[5]

A Joint Reserve Skyray from NAS Olathe on 1 November 1964. This aircraft has not received the DayGlo markings that were applied to most other reserve Skyrays. (Roger Besecker via Nick Williams Collection)

VMF-113 and VMF-215 retired their F-6As in April 1965. At the conclusion of the final flight a retirement ceremony was held to mark the end of the Skyray's service. (U.S. Navy via Nick Williams Collection)

The end of the road for VMF-113 and VMF-215 Skyrays came as late as April 1965 when much newer and higher-performance Navy jets like the supersonic Vought F8U Crusader and Mach 2 McDonnell F-4 Phantom II were already in the Navy's inventory. Photo shows good detail of the canopy, ejection seat actuator handles, and unusual patterning of the aircraft's flat-black nose radome. (U.S. Navy via Nick Williams Collection)

Chapter Eight

Test and Research Centers

YF4D-1 BuNo 130747 was one of the first 12 Skyrays produced. The aircraft spent its entire career in test service, first with the Naval Air Test Center (NATC) and later with the Naval Ordnance Test Station (NOTS). There, it was used in a highly classified secret program to attempt to launch America's first earth-orbiting space satellite. (Photo Courtesy Mark Nankivil)

Their names are legendary: Muroc, Farnborough, Edwards, and Domodedovo, just to name a few. These are all famous flight-test centers located in various countries around the world, both past and present. For the United States Navy, however, the name that said "flight test" more than any other was Pax River. Other Navair test centers specialized in weapons or systems testing, but their mission objective was always the same—give operational Navy pilots the very best aircraft they could possibly fly.

Naval Air Test Centers

The Naval Air Test Center (NATC), Patuxent River, Maryland, is the Navy's primary aeronautical testing facility. During the evolution of the Skyray the Test Center was organized into four divisions: 1) Flight Test, which was responsible for aircraft stability and control, engine performance, and carrier suitability; 2) Service Test, which evaluated operational and tactical suitability, emphasizing maintenance; 3) Electronics Test, which was concerned with avionic equipment; and 4) Armament Test, which evaluated the airplane as a weapons system. The Test Center became involved in the earliest stages of the Skyray program when, in July 1952, the first prototype XF4D-1 (BuNo 124586) was evaluated by a team of NATC pilots. These periodic evaluations by Navy pilots, known as Naval Preliminary Evaluations (NPEs), are conducted to verify the progress of the manufacturer in achieving flight

In March 1955 NATC received its first assigned Skyray, BuNo 130746. Painted in Sea Blue, it was one of the first 12 production airplanes. This Skyray was used in XC-7 steam catapult tests at Patuxent River. (U.S. Navy via Nick Williams Collection)

milestones, or to confirm modifications that were requested from earlier evaluations. NATC assumed custody of the XF4D-1s on at least two occasions, concluding with the Carrier Suitability Trials in July 1953.

Early production F4D-1s were subjected to NPEs by the Test Center in November 1954, and again in June 1955. NATC received its first permanently assigned Skyray (BuNo 130746, the seventh production article) in March 1955. By 1962, 23 more Skyrays would be assigned. NATC made extensive use of its large Skyray inventory, examining every aspect of their operation. The F4D-1s flew performance tests, flight characteristic tests, catapult and arrestment tests, barrier engagement tests, and external stores tests. They also flew target tow, chase aircraft, and drone-control missions during their tenure at Patuxent River. The most formal and prolonged test project conducted at NATC was the Board of Inspection and Survey (BIS) trials, which began on 19 November 1956 and concluded on 6 January 1958. The BIS trials (a process that dates back to 1868 when it was used to insure the seaworthiness and suitability for service of warships) was an overall determination of whether BuAer was getting the aircraft it bargained for.[1]

The Test Center Skyrays wore highly diverse color schemes. The earliest were painted in Sea Blue with *NATC* painted in white on the fin. Some Skyrays retained the Gull Gray and white scheme they had worn previously in the Douglas Testing Division. The most colorful NATC Skyray, BuNo 130743, was painted in the drone controller scheme consisting of an Engine Gray fuselage with yellow wings and a Fluorescent Red-Orange tail.

In 1960, as F4D-1 test programs were winding down, NATC gradually began to relinquish their Skyrays. The Test Center's last F4D-1, BuNo 134748, was retired in 1965.[2]

A rare look at the underside of 130747 during a landing approach. The insignia blue scheme used a block-lettered* NAVY *on the lower left wing, which did not appear on the later Gull Gray and white scheme. (Boeing via Nick Williams Collection)

On 14 April 1955 a second Skyray, 130747, was accepted at NATC and was employed in several test roles. It was transferred to NOTS, China Lake, California, on 11 August 1956. It was reassigned to NATC on 1 October 1959, and served until 7 November 1962 when it was stricken with 695 hours. (U.S. Navy via Nick Williams Collection)

BuNo 130743 on display at Andrews AFB, Maryland, 14 May 1960 painted in the drone controller scheme. This was the only Skyray to be painted in this scheme. BuNo 743 was used to tow high-speed dart targets. (Top: Stephen Miller; middle and bottom left: David Lucabaugh via Ron Piccianni Collection; middle and bottom right: Don Spering AIR Collection)

By the early 1960s NATC Skyrays received liberal doses of DayGlo paint. (Nick Williams Collection)

Naval Air Facility, China Lake

The Naval Air Facility (NAF) China Lake conducted all flight testing for ordnance developed by the U.S. Naval Ordnance Test Station (NOTS). To perform the flight testing of ordnance the NAF maintained a fleet of approximately 20 different aircraft types. From 1956 to 1961 Skyrays played an important role in these test programs at China Lake. The first Skyray (BuNo 130747) was assigned to China Lake on 11 August 1956. This aircraft was originally part of the Douglas Testing Division fleet, and then served at the NATC Patuxent River, Maryland. Four additional Skyrays joined 130747 at China Lake: 130745, 134762, 134765, and 134853.

NOTS was noted for developing the 5-inch High-Velocity Aircraft Rocket (HVAR), the Tiny Tim, the Mighty Mouse, the Zuni, and the Shrike, but its most notable development was the Sidewinder missile. The Sidewinder began in 1946 as a NOTS in-house research project on heat-homing rockets headed by William B. McLean. It received official funding from the Navy Bureau of Ordnance in 1951 and was given full program status in 1952. The prototype Sidewinder was successfully fired in September 1953 and it entered operational service in 1956. The Sidewinder evolved into the most widely used missile in the West with more than 110,000 examples produced for the United States and 27 allies, and has been in production for more than 50 years. The Skyray was instrumental in the success of the Sidewinder by serving as one of its first launch platforms in test and operational service.

During the development of the Sidewinder it became apparent that high-speed aerial targets were needed for air-testing ordnance. High-performance pilotless aircraft did not exist in the mid 1950s. The only drones at that time were converted World War II aircraft or first-generation jets. A China Lake team led by the Naval Air Facility's chief project pilot, LCDR R. R. "Van" Vancil, and Leonard W. Seeley, head of the Flight Evaluation Branch, devised a simple cost-effective solution: a plywood target in the shape of a dart that could be towed at high speeds. Towed targets (banners) had been used for decades, but at limited speeds. Furthermore, getting the target aloft without damaging it presented a problem.

The China Lake team assembled a plywood dart 12 feet long and 4 feet wide that was towed from its center of gravity by a nylon line that was strung between two 10-foot-high poles. The towing aircraft hooked the line with its tail hook and pitched up into a 30-degree climb forming a long drooping arc in the tow line that pulled the dart gently into the air. LCDR Joel Premselaar, flying an F4D-1 Skyray, proved that the concept worked perfectly. The target could be towed at speeds in excess of 400 knots, more than twice the speed of other towed targets.[3]

This is the first production F4D-1, BuNo 130740, photographed at the Naval Ordnance Test Station (NOTS), China Lake, California, on 20 October 1954. This Skyray was part of the Testing Division Fleet (Douglas) and is shown carrying four rocket pods for early ordnance tests. Six months later this Skyray would be damaged beyond repair in a violent transonic pitch-up. (U.S. Navy via Nick Williams Collection)

A Douglas Testing Division Skyray, BuNo 134762, undergoing tests on the Aero X24 rocket pod at China Lake on 9 November 1956. This Skyray is part of the second production batch, as evidenced by the enlarged inlets and boundary layer ramp. (U.S. Navy via Author's Collection)

This is the first Skyray assigned to the Naval Air Facility at China Lake, BuNo 130747, photographed on 18 May 1957. This Skyray had served in carrier suitability tests at the Naval Air Test Center, Patuxent River, Maryland, prior to its acceptance by China Lake. (Doug Olson via Bill Larkins)

In 1958, LCDR Premselaar along with LCDR William W. "Bill" West and CDR Hal Lang were involved in another notable China Lake Skyray project—a highly secret program known officially as Project Pilot (unofficially called NOTSNIK). On 4 October 1957, the Soviet Union successfully launched the first earth-orbit satellite, Sputnik. The United States was stunned by the Soviet success and looked for a means to respond rapidly.

The Naval Ordnance Test Station was one of the organizations that the government looked to because of its experience in developing air-launched weapons with solid fuel rockets. In short order NOTS designed a vehicle that relied on existing rocket motors along with other off-the-shelf components. It was designated NOTS-EV-1 and was to be carried by a Skyray to 41,000 feet where it would be released and the first of five stages would be fired. The five-stage burn was expected to place a 2.3-lb payload into an orbit 1,400 miles above the earth. The project was classified top secret and was shielded from the public for many years. There were six attempts to launch the NOTS-EV-1 from a Skyray. Four clearly failed when the vehicle exploded shortly after separation or when the first stage failed to ignite. However, two launches resulted in the vehicle climbing rapidly beyond the view of both the launch aircraft and the chase aircraft. There is some evidence that one of these launches did achieve orbit. A tracking station in Christchurch, New Zealand, reported a very weak signal at the times expected for the first and third orbits, but the evidence was inconclusive.[4]

The F4D-1 was used at China Lake to develop high-speed tow targets. LCDR Joel Premselaar perfected a method of picking up the target, as shown in these photographs. (U.S. Navy via Joel Premselaar)

Still wearing Douglas Testing Division markings, this Skyray, BuNo 130745, was assigned to the Naval Air Facility at China Lake. Shown here in April 1957 when it was being used for Sidewinder missile testing. Notice the Sidewinder logo on the nose. (U.S. Navy via Author's Collection)

This China Lake Skyray, BuNo 134762, has been modified to carry a high-speed dart target from its drop-tank pylon. Notice that the wingtips of 762 have embedded movie cameras. (U.S. Navy via Author's Collection)

A test store is mounted on the right mid pylon of 747 on 5 June 1958 to verify the Skyray's ability to carry this load to 41,000 feet prior to a live launch of a NOTS-EV-1 during Project Pilot. (U.S. Navy via author's collection)

The chase pilot for Skyray 747 sits in the cockpit of an F4J Fury. Two movie cameras are mounted on the sides of his helmet to film the Skyray's performance during the test flight. (U.S. Navy via Author's Collection)

A live NOTS-EV-1 is mounted on the left mid pylon of 747 in preparation for a satellite launch attempt on 19 July 1958. The NOTS-EV-1 was hastily made from off-the-shelf rockets and parts at China Lake. (U.S. Navy via Author's Collection)

LCDR Joel Premselaar boards 747 at dawn for a satellite launch. During Project Pilot six launches were attempted. There is some speculation that one launch may have successfully placed the 2.3-pound payload into orbit. (U.S. Navy via Author's Collection)

The last Project Pilot launch was flown by BuNo 130745. Here the NOTS-EV-1 launch vehicle is being inspected by LCDR Bill West on 5 August 1958. (U.S. Navy via Author's Collection)

Project Pilot was canceled after the sixth launch, but a second program known officially as Project Caleb (unofficially called NOTSNIK II) was approved in 1959. The orbital vehicle was designated NOTS-EV-2 and it was powered by four newly designed rocket motors. The first launch took place on 28 July 1960. It was a successful test of the first stage, which was carried aloft by the same Skyray that was used in the earlier program, BuNo 130747. The next launch was attempted in October, but the second-stage rocket failed to ignite, and Project Caleb was canceled.

Skyray 130747 was delivered to China Lake in a glossy Sea Blue scheme with a cream-colored radome and *China Lake* painted in white block letters on the tail. During the NOTSNIK project the underside of

A second satellite launch program was approved at China Lake in 1959 called Project Caleb. The launch vehicle was a higher thrust missile designated NOTS-EV-2. This pre-dawn photograph shows the Skyray and missile being prepared for flight on 28 July 1960 when the first stage was successfully fired. (U.S. Navy via Author's Collection)

By the summer of 1960 several of the China Lake Skyrays were repainted in the split scheme. The tail, nose, and leading half of the wing were Fluorescent Red-Orange; the remainder of the aircraft was gloss white. (U.S. Navy via Author's Collection)

the wings were painted DayGlo Orange, and the trimmers were painted white. The other Skyrays wore the gray and white scheme with red trim of the Douglas Testing Division. By 1960 several of the F4Ds were repainted in the high visibility split scheme, which consisted of a glossy white base with large swatches of Fluorescent Red-Orange.

In late 1960 all Skyray programs at China Lake were at an end, and its last F4D-1, BuNo 130745, was stricken with only 137 flight hours.[5]

Naval Missile Center Skyray 134856 was used to test the Sparro-air vehicle, which was powered by two Sparrow air-to-air rocket motors. The Sparro-air succeeded in reaching an altitude of 64 miles. (U.S. Navy via Nick Williams Collection)

Naval Missile Center

Not far from the Naval Air Facility China Lake is the Naval Missile Center at Point Mugu, California. The Missile Center received its first Skyray, BuNo 134752, from the Douglas Testing Division in February 1958. Over the next four years the Missile Center received four additional Skyrays: BuNos 130748, 134763, 134856, and 134875. These F4D-1s were used as missile-launch platforms, target tugs, and chase aircraft. In 1961, BuNo 134856 launched the first of 10 Sparro-air unguided sounding rockets. The Sparro-air was powered by two Sparrow air-to-air rocket motors and it succeeded it reaching an altitude of 64 miles.

The Point Mugu Skyrays all wore *Naval Missile Center* painted in large block letters on their tails, and all carried large amounts of Fluorescent Red-Orange. BuNo 134856 was the last Skyray to remain in service at the Missile Center. It was retired in 1962.[6]

Naval Air Test Facility

Established on 1 October 1957 the Naval Air Test Facility (NATF) was located at the historic Lakehurst Naval Air Station, a former Navy blimp field and the site of the Hindenburg disaster. The NATF was assigned the mission of research, development, testing, and evaluation of shipboard and shore-based catapult and arresting systems. The NATF used 11 test sites at Lakehurst, including

Skyray 134856 parked on the flight line at the Naval Missile Center, Point Mugu, California. This was the last Skyray to remain in service at Point Mugu. It was retired in 1962. (Piccianni Aircraft Slides)

Skyray 134875 was also assigned to the Naval Missile Test Center. This aircraft still wears its tactical Gull Gray and white scheme with DayGlo trim. (Piccianni Aircraft Slides)

Skyray 134854 was one of two F4Ds assigned to the Naval Air Test Facility, Lakehurst, New Jersey. BuNo 134854 is shown here in its original Gull Gray and white scheme. The chevron on the tail is International Orange and white with black stripes. (Nick Williams Collection)

The Lakehurst Skyrays were used for catapult tests from 1959 through 1964. This photograph captures 134854 being launched from an experimental steam catapult. (Nick Williams Collection)

In the early 1960s BuNo 134854 was repainted in the split scheme of gloss white and Fluorescent Red-Orange. The aircraft is pictured here during XRE-1 Expeditionary Catapult System tests in 1963. (Nick Williams Collection)

the longest military runway on the East Coast. A small fleet of contemporary fighter and attack aircraft, which included two F4Ds (134854 and 134931), were employed by the NATF, and were used from 1959 until 1964.

On 11 December 1959 NATF tested the Skyray's compatibility with the MK-7 Mod 2-3 arresting gear system (shipboard barricade). A total of 63 arrestments were made with this system, and the Skyray was found to be compatible, with minor limitations. From 16 May to 31 July 1960 an NATF Skyray was launched 19 times during an evaluation of the TC13 steam catapult. At the same time the second Skyray was employed in making 46 arrestments in an evaluation of the M4-1 Expeditionary Arresting Gear located at Patuxent River, Maryland. In June and July 1961 the TM-20 Expeditionary Arresting Gear was tested by an NATF Skyray. Throughout 1962 both Skyrays (now designated F-6As) were used to test the XRE-1 Expeditionary Catapult System, molybdenum-coated arresting hook points, and the MK-7 Mod-1 arresting gear system.

The last NATF test assignment for the Skyray was a demonstration of compatibility with the E-27 emergency arresting gear. Eight arrestments were made in August 1963. Both F-6A Skyrays were stricken from NATF's inventory in 1965.

The Test Facility Skyrays were originally painted standard Gull Gray and white with *NATF* painted in block letters on the tail. An orange chevron outlined with black stripes was painted at the top of the tail, and a nose number was painted aft of the radome. By 1963 BuNo 134854 was repainted in the high-visibility Split Scheme of glossy white with a Fluorescent Red-Orange nose, lower fuselage, tail, and front half of the wing.[7]

Naval Air Development Unit

The Naval Air Development Unit (NADU), commissioned on 1 September 1953, was located in the old blimp hangar at NAS South Weymouth, Massachusetts. Its primary mission was to evaluate the effectiveness of experimental and prototype detection, tracking, and intercept functions in air-defense systems. NADU employed numerous aircraft types ranging from a ZPG airship to an F4D-1 Skyray. Five Skyrays were assigned to the Unit; the first, BuNo 134825, arrived in August 1956. The others, 130748, 134748, 134930, and 134938, were received within the next two years. By 1962 NADU ended its use of F4D-1s, which were replaced by newer fighters.

The high-visibility Split Scheme of glossy white and DayGlo was used on some of the Unit's Skyrays with the block letters *NADU* painted on the tail and a nose number aft of the radome.[8]

A rear 3/4-view of BuNo 134854 lifting off from the Lakehurst runway. This view shows the upper wing markings of the split scheme. (Nick Williams Collection)

The F4D-1 Skyray from the Naval Air Development Unit test squadron based at NAS South Weymouth, Massachusetts, is on display at an open house inside the massive Navy blimp hangar located at that facility and also used by a Navy ZPG airship. Note tail of McDonnell F3H Demon at extreme right. (Nick Williams Collection)

Naval Air Development Center

Established on 1 August 1949 at NAS Johnsville, Pennsylvania, the Naval Air Development Center (NADC) was assigned the mission of providing design and development of aircraft electronics, aviation armament, and pilotless aircraft. By 1956, the Air Development Center had grown to include seven laboratories and was a major employer in the Philadelphia area. Two Skyrays were attached to NADC, BuNo 134763 and BuNo 134966. The Center's first Skyray, 134763, was transferred from the Douglas Testing Division in August 1957 and was employed by NADC until September 1959.[9]

This Skyray was operated by Marine Headquarters and Maintenance Squadrons 15 and 24 in support of the tactical missions of their parent Marine Air Groups (MAGs). (Clay Jansson via Nick Williams Collection)

H&MS-15 and H&MS-24

Headquarters and Maintenance Squadrons (H&MSs) provided training, administrative, and logistical support to their parent Marine Air Groups (MAGs). They often accompanied the MAG tactical squadrons on weapons training deployments. The H&MSs were expected to operate at least one example of every type of aircraft flown by the Marine Air Group. While it appears that several H&MS units had custody of F4D-1 Skyrays (such as H&MS-11 at NAS Atsugi, Japan, and H&MS-30 at Kaneohe, Hawaii) only H&MS-15 at MCAS El Toro, California, and H&MS-24 at MCAS Cherry Point, North Carolina, flew the airplane in significant numbers. H&MS-15 operated four Skyrays (134795, 134804, 134851, and 139111) during 1959. The tail code YV appeared on these El Toro Skyrays. H&MS-24 operated three Skyrays (134789, 134823, and 139085) with the tail code EW. These Skyrays were operated in 1961 and 1962.[10]

A family portrait of the Douglas aircraft undergoing experimental flight test in 1954. In the foreground is the X-3, on the left is the D-558-1, in the center is the XF4D-1, and on the right is the D-558-2. (NASA Dryden Flight Research Center Photo Collection)

NACA/NASA

The National Advisory Committee for Aeronautics (NACA) was one of the civilian agencies that operated the F4D-1 Skyray. Because of its unique configuration, NACA's Ames Research Center at NAS Moffett Field, California, acquired 134759 on 4 April 1956. This was one of the first production deliveries occurring just as the Skyray was entering the Fleet. NACA used the Skyray to perform a detailed evaluation of low-speed handling characteristics, high-altitude roll coupling tendencies, and transonic trim changes. In October 1958 NACA was reorganized into the National Aeronautics and Space Administration

One of the first production Skyray deliveries was 134759 to the NACA Ames Research Center at NAS Moffett Field, California, in April 1956. Here the NACA acceptance pilot is inspecting the aircraft. (Boeing via Nick Williams Collection)

A NACA pilot is briefed before research flight (left). NACA often flew these missions without external fuel tanks to obtain maximum performance. The yellow tail band, which encloses the NACA logo, can be seen (right). (NACA via Nick Williams Collection)

Later in its service with NACA BuNo 134759 received enhanced markings. **AMES** ***was painted on the fin, and*** **TEST** ***was painted on the upper and lower blended wing fillets. A long pitot boom has been added for data gathering.*** (NACA via Nick Williams Collection)

In 1958 NACA became NASA. BuNo 134759 was repainted in this DayGlo scheme, which retained the yellow tail band and **TEST** ***markings.*** (Bill Larkins)

(NASA). Skyray 759 continued in service with NASA until 16 October 1959 when it was transferred to the Naval Air Test Center at Patuxent River, Maryland, for additional test assignments.

Initially, 134759 was painted in the standard Gull Gray and white scheme. It carried a yellow tail band with black borders with the NACA winged insignia and the word *Ames* on the tail. For test purposes this Skyray had a long pitot probe mounted to its nose, and it had the word *TEST* painted in large block letters on the upper and lower wing fillet surfaces. Under NASA the aircraft was repainted in a vivid DayGlo scheme. It retained the yellow tail band with *NASA* in it.[11]

U.S. Naval Test Pilot School

The last unit to fly the F4D-1 Skyray was the U.S. Naval Test Pilot School (TPS), located at NAS Patuxent River, Maryland. The Test Pilot School began as an informal training program of the Flight Test Division at the Naval Air Test Center in 1945. By 1948 the program had evolved into a formal six-month course, and in 1958 it grew into an eight-month course with the school being officially named the U.S. Naval Test Pilot School. The school was charged with the mission of training experienced aviators, Naval Flight Officers, and engineers to become qualified engineering

Miscellaneous Units

The Fleet All Weather Training Unit Atlantic (FAWTULANT) had the primary mission of training replacement pilots for duty with Fleet units. In September 1957 Skyray 134861 was assigned to FAWTULANT at NAS Key West, Florida, followed by 134860 in October 1957, and 139050 in February 1958. Ultimately, this unit employed five Skyrays in the execution of its training duties.

Several organizations had custody of individual F4D-1s. Utility Squadron Seven (VU-7) at NAAS Brown Field, California, operated a Skyray (134772) on loan from neighboring VF(AW)-3 at North Island from June to August 1959. The Naval Aviation Ordnance Test Station (NAOTS) at Chincoteague, Virginia, flew Skyray 134754 for more than 100 hours from June 1957 until March 1959 before transferring the airplane to NATC Patuxent River, Maryland. The Inspector of Naval Material (INM), Boston, Massachusetts, operated Skyray 134748 from October 1957 to February 1961. Custody was transferred to the Bureau of Naval Weapons Representative (BWR), Waltham, Massachusetts, who bailed it to the Raytheon Company for flight testing of their Sparrow III air-to-air missile until its retirement in October 1963. Finally, the Navy Liaison Officer, Edwards Air Force Base, California, was assigned an individual Skyray, BuNo 134769, from August 1956 until April 1958.[12]

The Naval Aviation Ordnance Test Station (NAOTS), Chincoteague, Virginia, operated Skyray 134754 from 1957 to 1959. BuNo 134754 is seen here performing a Dart target pick-up on 28 August 1958. The maneuver was developed one year earlier at the Naval Ordnance Test Station, China Lake, California. (U.S. Navy via Nick Williams Collection)

Still in its Douglas Testing Division colors 134769 is seen at the Naval Air Development Unit (NADU) hangar, South Weymouth, Massachusetts, on 17 September 1961. At that time 769 was being operated by the Raytheon Company to flight test the Sparrow III air-to-air missile. (Jim Burridge via Nick Williams Collection)

The first Skyray assigned to the U.S. Naval Test Pilot School (TPS) was 134764 in November 1961. The School operated a total of three Skyrays until 1969. They were used to provide tailless delta experience for the students. (Stephen Miller; Piccianni Aircraft Slides; Bernard Schulte via Nick Williams Collection)

test pilots, test flight officers, and test project engineers. The school's graduates went on to serve at the Naval Air Test Center, other Navy research and development facilities, as well as testing centers of other U.S. and allied military services.

In the 1960s the course was divided into academic and flight training. During the eight-month course each student would average 1 flight hour and 3½ classroom hours each day. Academic study included aerodynamics, thermodynamics, dynamic and static stability, jet engine theory, and weapons systems. Flight training prepared the student to plan, fly, and write a test flight report on various assignments dealing with aircraft performance or stability and control.

TPS maintained a very diverse fleet to expose its students to a broad array of aircraft. A student could be assigned to an F8U Crusader or A4D Skyhawk one day, and a T-28 Trojan or Schweitzer 2-32 sailplane the next. The F4D-1 Skyray joined the TPS fleet toward the end of its front-line service to provide a tailless delta-wing example. The first Skyray, BuNo 134764, was assigned to the school in November 1961. By the mid 1960s two additional Skyrays, BuNos 134806 and 139177, were acquired.

According to LtCol Philip Oestreicher, USMCR (Ret.), a former VMF(AW)-114 Skyray pilot, and TPS student, "When I attended the school in 1966, the F4D-1 (actually F-6A then) was used to demonstrate unusual flying qualities—in modern combat airplanes—such as high adverse yaw, high dihedral effect, and rather low directional stability. My study group used it as a performance testing subject only... A member of the class before mine got one into a spin and was unable to recover. He ejected successfully."[13]

BuNo 134806 parked on the flight line at the U.S. Naval Test Pilot School displays its DayGlo markings. (Nick Williams Collection)

In 1968 the Test Pilot School changed the tail markings on BuNo 134806 from TPS to a stylized Skyray image. (Nick Williams Collection)

The last Skyray acquired by TPS was 139177. Initially this Skyray had a book with the letters TPS painted on its tail. (Phil Oestreicher)

The three Skyrays assigned to the Test Pilot School were painted in a similar scheme. The standard Gull Gray and white was given a DayGlo Red-Orange tail, nose, and wingtips. The letters *TPS* were painted on the tail (the design varied among the three aircraft), and *U.S. Naval Test Pilot School* was painted on the spine.

This F4D-1 Skyray calls the U.S. Naval Test Pilot School its home. Shown parked next to a Lockheed P-3 Orion patrol aircraft, this Skyray sports an understated, yet handsome, color scheme accented by bands of high-visibility DayGlo Red-Orange paint. Note extended outboard leading-edge slats. (Nick Williams Collection)

The F4D-1 with its high performance and quirky behavior spent its final flying days as a demonstration tool for the test pilot community. By the late 1960s, it became economically infeasible for the school to continue flying Skyrays. On 25 November 1969, Capt. George Watkins ferried the last airworthy F-6A, BuNo 134806, on a 2.3-hour flight to the National Naval Aviation Museum at Pensacola, Florida. The remaining Skyray airframes, BuNos 139177 and 134764, were placed on display at the Marine Corps Air Museum, Quantico, Virginia, and the Naval Air Test and Evaluation Museum, at Patuxent River, Maryland. The Test Pilot School replaced their Skyrays with Northrop T-38 Talons, which were still in use as of 2009.[14]

F-102A Operational Service

The first operational F-102A was delivered to the 327th Fighter Interceptor Squadron at George Air Force Base on 24 April 1956—only eight days after the first F4D-1 was delivered to VC-3 at Moffett Field. By the end of the year, 97 F-102As had been delivered to five Air Defense Command (ADC) squadrons. In 1957 16 additional Air Defense squadrons converted to the Delta Dagger (nicknamed the Deuce) bringing the total number of delivered aircraft to 428—more than the entire Skyray production run. By the end of 1958, 27 ADC interceptor squadrons were flying 651 F-102As. However, many of these aircraft were not actually combat ready. Like the Skyray, the F-102A experienced many teething problems. These ranged from exploding engine starters to landing-gear problems. The armament was also failure prone, and an Air Force synopsis concluded that the F-102 weapons system did not provide the ADC with an acceptable combat capability.[15]

In 1959 European and Pacific commands began acquiring F-102As from the Air Defense

In 1958, the 317th Fighter Interceptor Squadron moved to Elmendorf Air Force Base in Alaska to provide a first line of defense against Soviet aircraft that made a habit of probing Alaskan airspace. (U.S. Air Force via Don Spering/AIR Collection)

Command. The ADC's use of the interceptor contracted further as F-102As were pulled from its inventory to meet commitments for the war in Southeast Asia. By 1968 only one Fighter Inceptor Squadron in the ADC used the F-102A and by 1973 it was fully retired from that command.[16]

The F-102A was not well suited for use in Southeast Asia. It was first deployed in 1961 in an air-defense role, but its role was expanded to include air-to-ground operations against supply routes on the Ho Chi Minh Trail, day-fighter operations in close-air-support missions, and combat air-patrol escort missions for B-52 bombers over North Vietnam. By 1968, F-102A operations in Southeast Asia began to wind down, and by 1970 their activities were taken over by F-4 Phantom IIs.[17]

The F-102A enjoyed a long affiliation with the Air National Guard (ANG). Twenty-three ANG units operated the Deuce. The first was the 182nd Fighter Interceptor Squadron at Kelly Air Force Base, Texas, on 4 June 1960. The last was the 199th Fighter Interceptor Squadron at Hickam Air Force Base, Hawaii, in January 1977.[18]

The F-102A also served with the Turkish and Greek air forces, but the Deuce's final role was as a full-scale target drone. In February 1972 nearly 400 F-102As were in storage. The Sperry Rand Corporation was awarded a contract on 1 April 1973 to modify F-102As into one of two drone configurations. QF-102As retained pilot controls for contractor-operated flights, while PQM-102As were configured for pilotless flights. These became the first full-scale targets with afterburner capability. When Sperry's production ceased in 1981, a total of 215 F-102As had been converted. The Deuce drone had a basic weight of 20,000 pounds with a mission gross weight of 31,200 pounds. It could carry a wide variety of electronic countermeasures and attain a speed of Mach 1.2 at 56,000 feet. It could withstand a force of 8 g, and its infrared signature was basically the same as several Soviet Bloc fighters. In sum, it was a highly realistic target and provided excellent aerial testing for several ordnance systems.[19]

The first operational F-102A was delivered to the Air Defense Command's 327th Fighter Interceptor Squadron at George Air Force Base, California, on 24 April 1956. (National Archives via Mike Machat)

A QF-102A drone with a pilot on board. Many of the Sperry drone conversions allowed the aircraft to be flown remotely or with a pilot. This photograph was taken during the 1982 William Tell interceptor competition at Tyndall Air Force Base, Florida. (Don Spering/AIR Collection)

In June 1958 the first F-102A squadron, the 327th FIS, was deployed to Thule, Greenland. This is a photograph of the first Deuce to land at Thule. Note the bare-metal auxiliary fuel tank and canopy coming open. (U.S. Air Force via Don Spering/AIR Collection)

The Air National Guard (ANG) had a long affiliation with the F-102A. This photograph of a California Air National Guard Deuce was taken on 7 September 1974. The ANG retired their F-102As in January 1977. (U.S. Air Force via Don Spering/AIR Collection)

This Delta Dagger, photographed in November 1974 at Suffolk County Air Force Base on Long Island, New York, was attached to the 102nd FIS of the New York Air National Guard. It is painted in the Southeast Asian camouflage scheme of tan with two shades of green and light gray undersurfaces. (A. Hafter via Don Spering/AIR Collection)

Chapter Nine

F5D-1 Skylancer

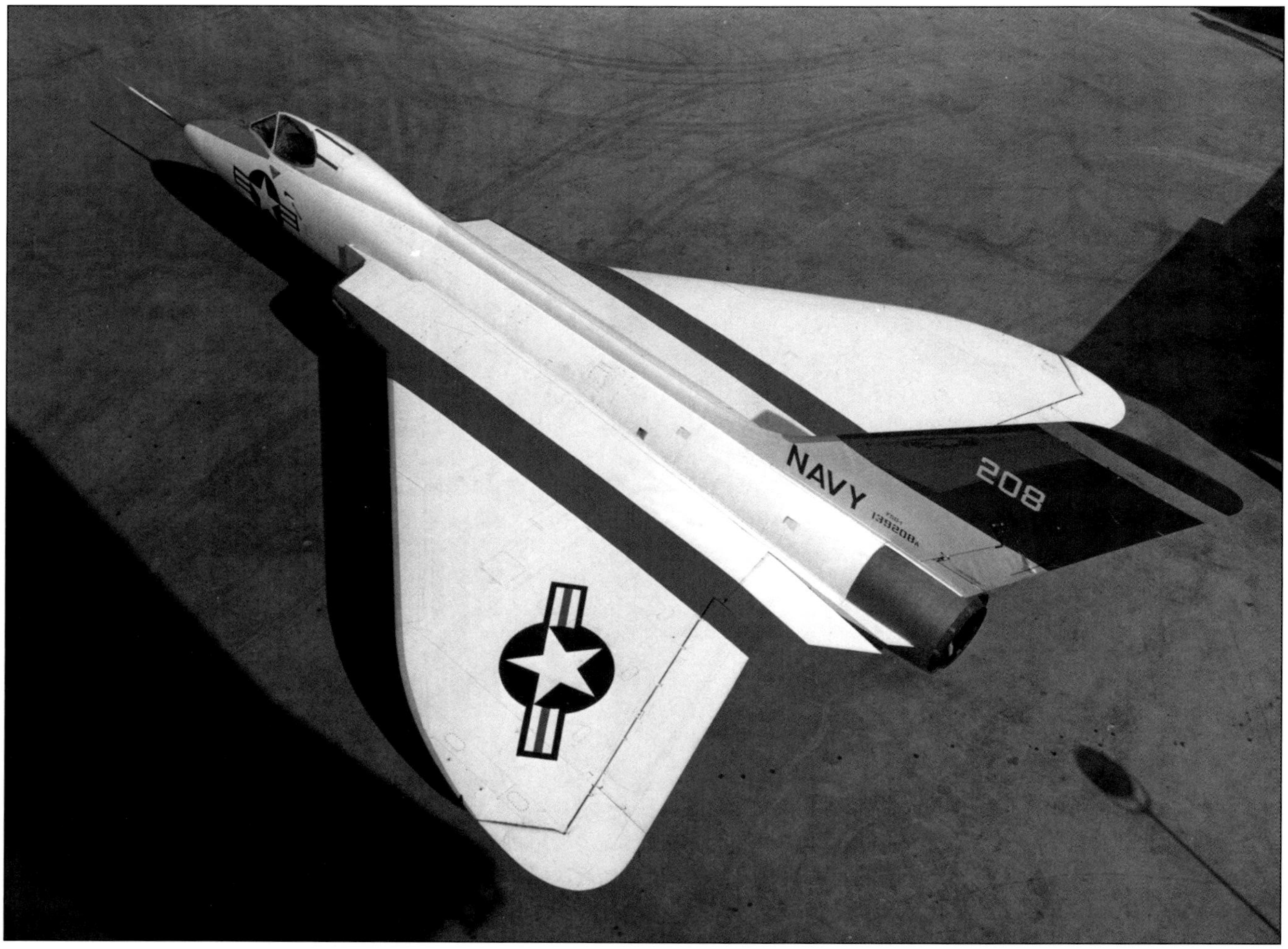

Looking every bit as futuristic as any sci-fi manta-ray secret weapon, the F5D-1 Skylancer prototype had originally been designated as the F4D-2, a practice quite common in the industry at that time. (Boeing via Tommy Thomason)

The Korean War was not only an international military conflict, it also served as a demarcation in aircraft design. Aircraft manufacturers worldwide were envisioning a new generation of larger, faster, and more powerful jet fighters. For the Douglas Aircraft Company, that airplane would become the F5D Skylancer.

Significant Changes

By late 1953, the Skyray had established itself as an exceptionally promising airplane. It had set two world speed records, passed its Carrier Suitability Tests, and earned its designer, Ed Heinemann, the highest award in American aviation. Yet it was apparent that the Skyray's basic engineering from the late 1940s was dated.

Douglas had little difficulty convincing the Navy to authorize work on an improved Skyray. It was to be designated the F4D-2N and it would incorporate the most recent advances in aerodynamics, structures, avionics, and ordnance. It would serve as an interceptor, general-purpose fighter, and ground-support bomber. It was to be armed primarily with 72 2-inch folding-fin aerial rockets (FFARs) carried internally in a retractable rocket launcher. The rocket package could be removed and replaced with a gun package of four 20mm MK 12 guns each carrying 125 rounds. The aircraft could also be equipped to carry two Sparrow I missiles on external pylons. The armament control system was to be the Aero 13G, an update of the Skyray's Aero 13F, consisting of Westinghouse APQ-50A radar with an MK 13 sight unit.

The forward fuselage of the Skylancer mock-up shows many of the aerodynamic improvements over the Skyray. The nose has been elongated, a V-shaped windshield is used, and the inlet ducts have been narrowed. The Skylancer was designated the F4D-2 at this time. (Boeing via Tommy Thomason)

BuAer issued a letter of intent dated 16 October 1953 for the construction of two airplanes and a static test article. A mock-up was completed and satisfactorily inspected on 10 November 1953. Design specifications were finalized on 9 February 1954, and work was started on the first flying article, which was assigned BuNo 139208.

The Skylancer, as it was named, represented a total re-engineering of the Skyray. Acknowledging that it was a totally new airplane with expanded mission requirements, the Navy chose to abandon the F4D-2N designation and call it the F5D-1. The fuselage was lengthened by 8 feet to improve its fineness ratio, thereby reducing drag and providing enough volume to double the internal fuel.

A rear 3/4-view of the F4D-2 mock-up. The increased fin height and the improved exhaust area are apparent. (Boeing via Tommy Thomason)

There has been conjecture over the use of Dr. Richard Whitcomb's Area Rule principles in the design of the Skylancer fuselage. Some Douglas public relations literature even makes reference to the Area Rule philosophy prevalent at the time. However, Douglas test pilot, Bob Rahn, noted in his biography, "The Air Force's F-106 delta-wing fighter was making its initial flight tests about this time, and it had the *Coke bottle* fuselage (pinched in) that was supposed to do wonders in reducing drag. I asked Leo Devlin (Ed Heinemann's assistant chief engineer) why he didn't do this for the

Technicians install a mock-up of the Pratt & Whitney J57 in the Skylancer fuselage to verify the fit for the mock-up review on 10 November 1953. (Boeing via Tommy Thomason)

The Instrument panel mock-up in the F4D-2 cockpit. The actual instrument panel would change substantially from this mock-up. (Boeing via Tommy Thomason)

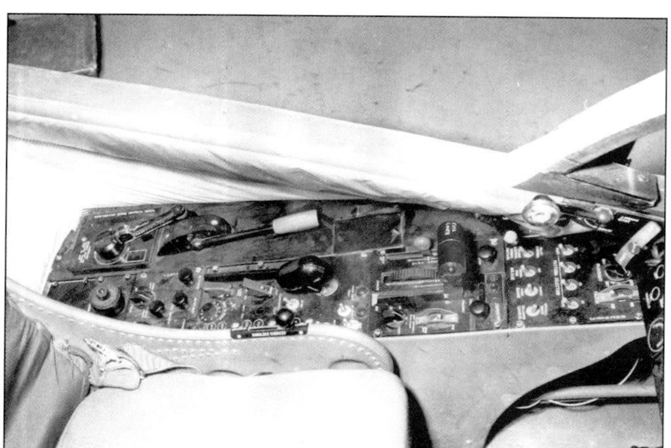

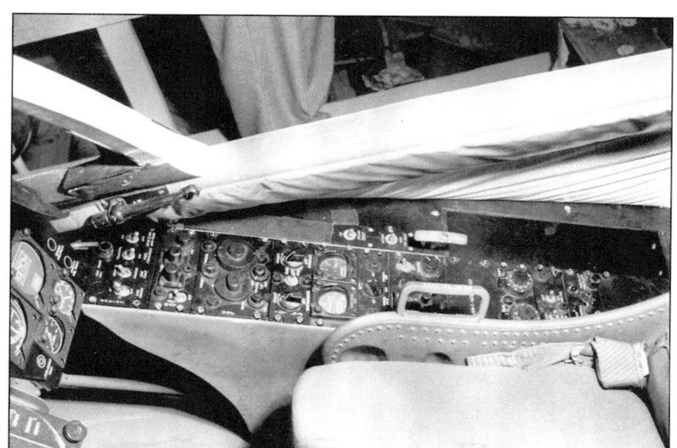

The left and right consoles in the F4D-2 mock-up. (Boeing via Tommy Thomason)

Skylancer. He replied, 'The fuselage is as thin as we can possibly make it, and it doesn't make sense to add bulges to the fuselage to get the Coke bottle effect.'"[1]

The Skyray wing, which was designed at the dawn of supersonic flight, was unnecessarily thick. Heinemann reduced the thickness of the Skylancer wing by 30 percent. However, the wing planform remained identical to the Skyray, providing 557 square feet of area. The wing fold for carrier storage was changed from the Skyray's mid-span position to a 3-foot section near the tip. It folded manually and locked mechanically, saving a significant amount of weight over the hydraulic system of the Skyray. The inlet ducts were flattened and moved forward on the fuselage with a wedge-shaped inlet ramp to avoid compressor stalls by bleeding off boundary layer air. The blended wing/body fillets were thinned, and the trimmers were modified to act as elevons in certain flight regimes. The canopy contours were changed, and a new V-shaped windshield similar to the F-102A was employed in place of the Skyray's flat plate windshield.

Heinemann was particularly eager to get rid of the thin skin surfaces of the F4D-1, which he now regarded as a mistake. The Skylancer used internally stiffened skins that were milled from light-alloy plate, nearly three times thicker than the Skyray's skin. Heinemann remarked in his biography, "New milling machines had been purchased, which featured horizontal spindles and the capability of machining large sheets of aluminum alloy. Skin thickness, compared to the Skyray's, was increased from 1/32 of an inch to 1/10. The machines permitted manufacture of integral stiffeners, and as a result, a foot-stamping stroll with hobnail boots across the Skylancer's extraordinarily smooth and strong surface would not be a matter for concern."[2]

The vertical tail was heightened by 1.8 feet and increased in area by 47.5 percent. The main gear tread was increased by 2.4 feet for additional ground stability. The Pratt & Whitney J57-P-8 was retained for use in the Skylancer. This provided only 16,000 pounds of maximum thrust for the enlarged airplane, but engine upgrades were expected and Heinemann made allowances for future engine growth. The Mock-Up Inspection Report noted, "It is expected that the improved Pratt & Whitney JT3C-20 [later designated J57-P-14], 17,200 pounds in afterburner and 10,900 pounds in military, will be available for fleet production F4D-2s in January 1957." The tail cone was extended beyond the trimmers, and a single-surface rudder replaced the Skyray's split rudder.

The first prototype F5D-1 at rollout displaying its wing-fold system. A 3-foot section of the wingtip folded manually and locked mechanically. This saved a considerable amount of weight and complexity. (Boeing via San Diego Air & Space Museum)

The first prototype, BuNo 139208, at rollout. By now the Skylancer had been redesignated the F5D-1. Its Skyray heritage is apparent, but it has numerous aerodynamic improvements. All F5D-1s were painted overall gloss white with red trim and only minor variations in markings. (Boeing via Tommy Thomason)

By September 1954, it became apparent that the government-supplied Aero 13G fire control system would not be available in time for the Skylancer test program. This system was essential in evaluating the Skylancer as a weapons system. BuAer requested a proposal from Douglas for a new fire control system. Douglas, serving as general contractor, designed a new fire control system from scratch that showed enormous promise. The radar was subcontracted to Westinghouse, the Avion Division of the American Car and Foundry Company developed the ballistic computer, and the A.C. Division of General Motors manufactured the compensating gunsight. Douglas developed the flight data computer and the autopilot, which used newly emerging solid-state electronics technology. The system was a dramatic advance in performance and reliability over the tube-based electronics of the Skyray.

On 18 November 1954 BuAer issued a letter of intent for development of the new Douglas system now designated the Aero X24A armament control system. BuAer authorized sufficient equipment for the system to be installed and tested in five Skyrays and two Skylancers. Aero X24A would permit the Skylancer to perform its expanded mission as an all-weather interceptor, general-purpose fighter, or ground-attack bomber. The cockpit components were carefully designed to provide the best possible display by reducing reflected light from the scope face and minimizing the pilot's angular eye accommodation. Minimum time between detecting a target and obtaining an actual lock-on was emphasized. All components were arranged for ease of maintenance and rapid "go" or "no-go" testing.[3]

BuNo 134763 was one of the five Douglas Testing Division F4D-1 Skyrays used to test the X24A weapon control system that was being developed for the F5D-1 Skylancer. This system promised significant improvements in accuracy and reliability by using new solid-state electronics. (Boeing via Nick Williams Collection)

XF5D-1 Test Program

Douglas pilot Bob Rahn, who flew most of the Skyray test program, was selected to make the first Skylancer flight. He recalled in his memoir, "The Skylancer was a marked improvement over the Skyray… It was the best plane I ever flew. The Navy called the F4D 'the Ford,' and I called the F5D, a 'Cadillac.'"[4]

On 21 April 1956, Rahn took the Skylancer aloft on its maiden flight from Edwards AFB. It performed well in all respects except for a low-amplitude lateral oscillation that was corrected by tightening the control system. The overall aerodynamic improvements in the Skylancer allowed it to exceed the speed of sound on its first flight, whereas the Skyray had required 18 months and numerous modifications before flying supersonically. The need for trim compensation in the transonic region was less than half that required for the Skyray, and the Skylancer could fly 63 percent faster than the Skyray with the same engine. Heinemann's aerodynamic refinements were providing dramatic benefits.

The emergency manual flight control system of the Skyray was eliminated and in its place a wind-driven hydraulic pump was installed. It could be extended into the slipstream to provide enough power to actuate the controls. The Navy wanted the Skylancer's role expanded beyond interceptor to also include fighter tasks. This meant that the Skylancer had to be more maneuverable than the Skyray at supersonic speeds. Rahn explained, "We achieved this by increasing elevon surface travel, increasing power in the hydraulic actuating cylinders, and hooking up the pitch trimmers to be used in conjunction with the elevons. We named these control surfaces the inboard and outboard elevons. With this arrangement the F5D could attain 5 Gs at supersonic speed—twice as much as the Skyray!"[5]

The Skylancer's new aerodynamics seemed to solve most of the Skyray's shortcomings, but Rahn discovered an unexpected problem. The lengthened fuselage resulted in some pitch instability, which diminished target tracking in the transonic region. This was the speed where most fighting was likely to occur. The problem was solved with the installation of a pitch damper that cost only 16 pounds in increased weight, but provided the added benefit of eliminating pitch-up when deploying the speed brakes.[6]

Rahn also discovered that the XF5D-1 had an unexpectedly low stall speed, "The stall program for the F5D was very limited, which was unusual for a prototype military airplane. The reason, however, was because a delta-wing design has a very flat lift coefficient at high angles-of-attack, with no discernible break in the curve. This made it difficult to determine when the airplane had reached a full stall.

"When I approached a stall while in the clean [gear up] configuration, power at idle, I established a rate of decrease in speed of not more than 1 knot per second. I wanted to precisely ascertain at which point the nose-down pitch would take place. The Skylancer had excellent stall-warning characteristics including general airframe buffet at 30 percent above the stall speed of 70 knots. I distinctly remember this number because of the unusual environment at the time: high alpha, high sink rate, moderate airframe buffet, low IAS, and full aft stick deflection."[7] The stall occurred with a mild yaw and roll to the right. Rahn was certain that the longer fuselage and larger fin would eliminate the need for the unconventional spin recovery techniques of the Skyray.

During the Skylancer's first month of testing it demonstrated an exceptional reliability for a prototype aircraft. It flew 21 times for a total of 24 hours. After three months it had completed 57 hours in 50 flights, which included three weeks of maintenance when no flight operations were conducted.[8]

The second prototype, BuNo 139209, at El Segundo in July 1956. This F5D-1 had sharper radii on its inlet lips and was used to explore high-altitude compressor stalls. The sharper inlet improved the Skylancer's top speed allowing it to reach Mach 1.63. (Boeing via Tommy Thomason)

An early test flight of the first prototype F5D-1 Skylancer with an F4D-1 chase plane in trail. The Skylancer's lengthened fuselage and heightened fin are apparent in this photograph. (Boeing via Nick Williams Collection)

The elegant lines of the Skyray are evident in this photograph as we see the first prototype flying in the hallowed airspace above Edwards Air Force Base in April 1956. Douglas test pilot Bob Rahn called the F5D "the best airplane I ever flew." (Boeing via Tommy Thomason)

The first two Skylancers in flight over Edwards during a Naval Preliminary Evaluation. Although their exhaust plumes are hard to detect in this photo, both aircraft are flying in full afterburner. (Boeing via Nick Williams Collection)

The second prototype, BuNo 139209, executes a carrier type approach at Edwards. The Skylancer proved to have excellent approach characteristics with minimal dispersions of touchdown points. (Boeing via Nick Williams Collection)

The second XF5D-1 (BuNo 139209) made its initial flight on 30 June 1956. Once again, Bob Rahn was the pilot. The inlet ducts on this Skylancer had been thinned for increased speed and it was feared that they might induce compressor stall. On 9 November 1956 Rahn flew to 54,000 feet and performed maneuvers at extreme angles-of-attack with heavy airframe buffeting. He encountered no compressor stalls—the inlet design was deemed a success.[9]

On 10 July 1956 five Navy test pilots headed by CDR Tom Gallagher arrived at the Douglas Test Facility to conduct the first Naval Preliminary Evaluation of the XF5D-1s. The evaluation consisted of 28 flights conducted over a seven-day period. Only minor discrepancies were noted and were quickly addressed by Douglas. The Skylancer development was going very smoothly, but that would soon change.

In late July 1956, Douglas revealed that the development cost for the Aero X24A fire control system was escalating rapidly. It was originally authorized as a $13.5 million program, but the costs had escalated to more than $34 million and they were expected to exceed $50 million. The Aero X24A system was a key feature of the F5D program, a state-of-the-art weapons-delivery system that was essential to the performance of the Skylancer's broad mission. More bad news followed when Douglas moved the Fleet delivery schedule from July to October 1958 and also determined that the combat radius and mission time would be 25 percent less than the contract specification. In addition, Douglas informed BuAer that the Aero X24A would be delayed for six months because of component availability problems.[10]

The first prototype Skylancer banks away from the chase plane to reveal its undersurface. The rocket bay doors and the one-piece main landing gear doors are notable differences from the Skyray. (Boeing via Tommy Thomason)

The entire fleet of F5D-1 Skylancers photographed on 15 June 1957 at El Segundo. The markings of the four aircraft were nearly identical with the exception of radome colors, tail stripeage, and nose art. The middle two aircraft, tail numbers 349 and 350, have wing pylons in place for Sparrow II missiles. (Boeing via Nick Williams Collection)

A second Naval Preliminary Evaluation was conducted from 26 March through 5 April 1957. CDR Tom Gallagher headed the NPE team again with LT Bill Lawrence and future astronaut, LCDR Alan Shepard. On the day the Navy test pilots arrived at Edwards, the first XF5D-1, BuNo 139208, caught fire and sustained serious damage to its lower fuselage, inlet ducts, and wing leading edges. Douglas worked around the clock to repair the aircraft and made it available for the evaluation; however, most of the flights were made by the second and third aircraft, 139209 and 142349.

Gallagher pronounced the Skylancer ready for production with minor modifications, but the Bureau of Aeronautics decided otherwise. Extreme budget pressure caused BuAer to fund only 11 aircraft. Eventually even that order was rescinded and the program was terminated with the completion of the fourth airplane.[11]

In an effort to save the Skylancer, Heinemann and Rahn flew to Washington to propose a Mach 2.0 version to the Secretary of the Navy. This Skylancer would be powered with the new General Electric J79 engine and incorporate variable-geometry air inlets. But the Navy had no budget for the Skylancer and a truly exceptional airplane was abandoned.

Rahn remarked, "Thus ended the F5D Skylancer testing by Douglas. Ed Heinemann believed, as did I, that the airplane was a genuine winner, which increased our disappointment in never seeing it go into production.

"The flight test program went so smoothly that there were only a few lessons learned:

1. All improvements are not exempt from causing discrepancies. For example:
   a. The use of inboard elevons for increased longitudinal control power, when differentially deflected for roll, caused perverse yaw, which resulted in unacceptably high roll rates.
   b. The lengthened fuselage for increased fineness ratio and additional fuel tanks resulted in increased pitch inertia and decreased static stability, and required a pitch damper.
   c. The new high force feel-spring installed to reduce longitudinal sensitivity resulted in high control forces at supersonic speed.
2. Deep stall investigations require a spin chute. Even though 47.5 percent more fin area was added, the high angle-of-attack required to attain maximum coefficient of lift for a delta wing, and the flat characteristics of the curve at this point, caused the airplane to spin inadvertently.
3. The customer is not always right. The Navy should have authorized production of the F5D Skylancer."[12]

Heinemann wrote that the Navy chose to fund the Vought F8U Crusader in lieu of his Skylancer because Vought had no other orders at the time, and Douglas was busy with five fully funded programs. Heinemann received some personal consolation, however, when he received a call from NACA asking why the Navy selected the F8U when, according to NACA tests, the F5D was .2 Mach faster. Heinemann, Rahn, and many others believed that the Skylancer was the superior airplane.[13]

NACA Use

On 20 August 1957, the first and fourth aircraft, BuNos 139208 and 142350, were flown to the National Advisory Committee for Aeronautics (NACA) Ames facility at Moffett Field for use in high-speed research. The second and third aircraft, BuNos 139209 and 142349, were bailed back to Douglas for use as high-speed chase and utility aircraft. In June 1958, the two bailed aircraft were also flown to Ames and turned over to NACA. There is no record of those Skylancers, 139209 and 142349, being used for NACA flight operations. They were most likely cannibalized for parts to keep the two airworthy Skylancers flying.

While at NACA, the Skylancers underwent numerous paint and markings changes. Initially, 139208 became NACA 212 and 142350 became NACA 213. On 1 October 1958, NACA changed its name to the National

Close-up of the NACA and Douglas Testing Division logos on the fourth Skylancer, which was now identified as NACA 213. (Nick Williams Collection)

In August 1957 the first (139208) and fourth (142350) Skylancers were delivered to NACA Ames Research Center at Moffett Field, California. The markings were little changed from the original Douglas Testing Division colors. The word TEST was painted in large block letters on the upper and lower wing fillets, and NACA logos were placed on the nose and upper fin. (Bill Larkins)

On 1 October 1958 NACA was replaced by a new agency, NASA. The F5D-1s received new markings to reflect the change. This is the first prototype, BuNo 139208, in NASA markings. This Skylancer received the tail number 212. (Bill Larkins)

NASA 213 in the early 1960s while it was still attached to Ames. The aircraft has been repainted with large areas of DayGlo. The large block-lettered TEST *on the upper and lower wing fillet would soon be removed.* (Peter Bowers Collection via Nick Williams)

In 1962 NASA 213 had been transferred to the Dryden Flight Research Center where it was used to develop an aborted launch maneuver for the Dyna-Soar Program. Neil Armstrong, who is pictured in the cockpit of 213, did much of the work. The block-lettered TEST has been removed from the Skylancer at this time. (NASA Dryden Flight Research Center Photo Collection)

To simulate the visibility from the cockpit of the Dyna-Soar capsule a transparent amber mask was applied to the F5D canopy. When the pilot lowered a blue tinted visor on his helmet the combination of the blue and amber tints made the masked portions of the canopy opaque. (NASA via Tony Landis)

This picture of Skylancer 213 on 1 September 1961 was taken shortly after the aircraft arrived at Dryden. The test boom has been removed and a camera has been installed in the radome. (NASA Dryden Flight Research Center Photo Collection)

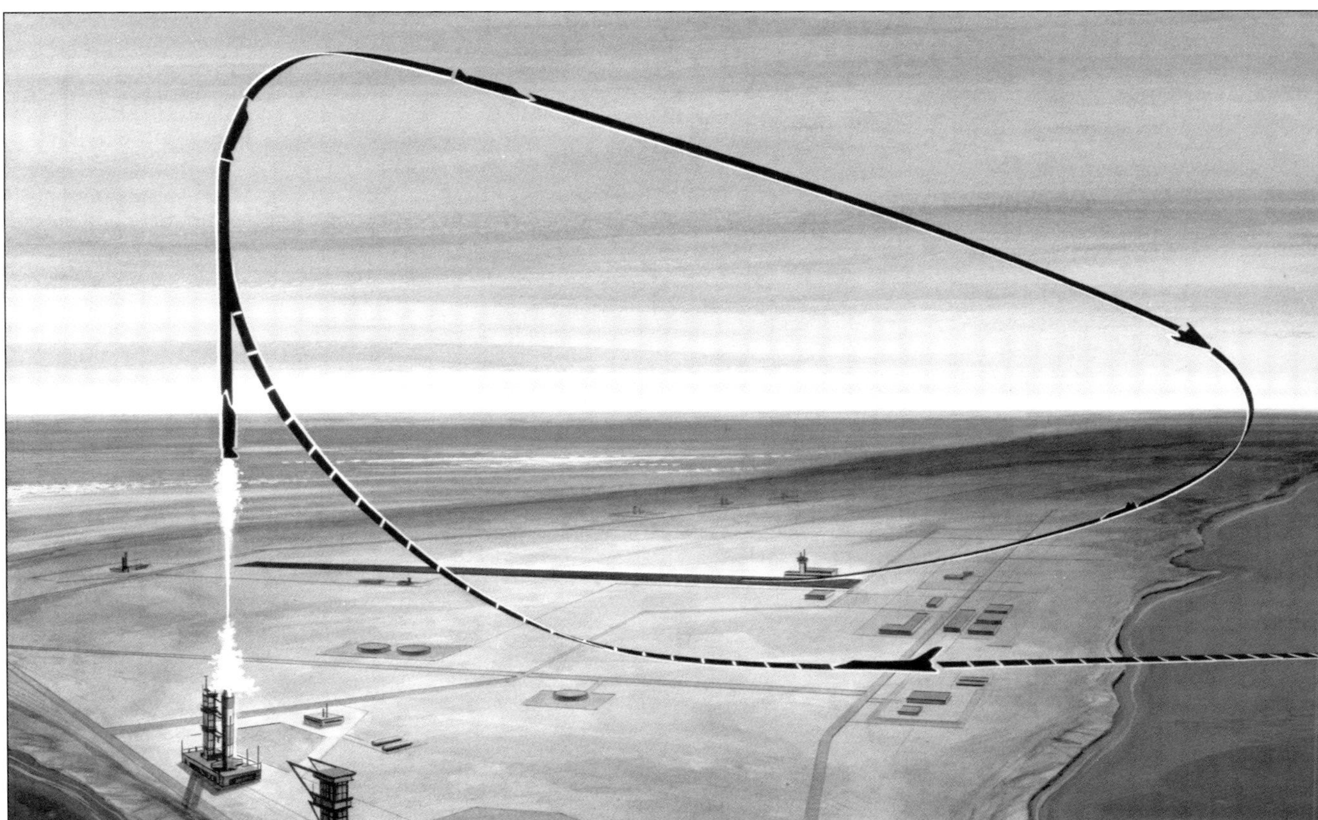

This illustration describes the path of the F5D during the simulated abort mission. (NASA via Tony Landis)

Aeronautics and Space Administration (NASA) and the Skylancers were remarked to NASA 212 and NASA 213. In 1961, both Skylancers were transferred to NASA Dryden at Edwards AFB. In 1963, Skylancer 139208 was returned to Ames at Moffett Field where it was renumbered NASA 708. It retained that identifier until it was retired from flight operations on 9 April 1968. BuNo 142350 remained at NASA Dryden until its retirement in July 1970. It had been renumbered NASA 802.

NASA used BuNo 139208 to investigate leading-edge vortex phenomena. It was discovered in the late 1950s that substantial aerodynamic lift could be realized from the vortex of air generated by a sharp wing leading edge. Unlike wingtip vortices, which flow outward

Skylancer 212 is the first prototype 139208. The aircraft was still attached to Ames Research Center at Moffett Field when this picture was taken. The aircraft is painted in DayGlo markings similar to 213. However, the nose's trim line cants forward, while nose trim on 213 cants aft. Note that 212 carries the large TEST marking at this time. (NASA Dryden Flight Research Center Photo Collection)

NASA 212 after it was transferred to Dryden in August 1961. The Ames markings have been removed from the tail, and the TEST markings have been removed from the fuselage. This Skylancer would soon have its wing modified to explore leading-edge vortex phenomena. (NASA Dryden Flight Research Center Photo Collection)

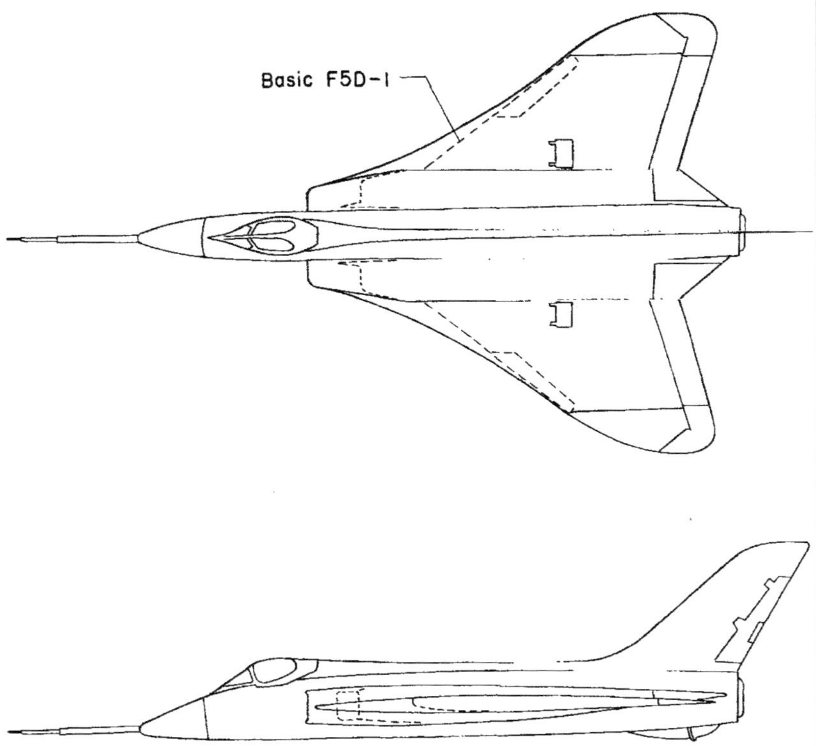

Figure 2.- Two-view sketch of test airplane.

A two-view drawing of the wing modification to NASA 212. The ogee curve leading edge was grafted onto the basic F5D-1 wing. It was built from plywood, fiberglass, and aluminum. The modification added 104 square feet of wing area, and allowed a 10-knot reduction in the landing approach speed. The wing modification required that the inlets be moved forward. (NASA Technical Note D-3071 Figure 2)

behind the wing and produce drag, leading-edge vortices extend spanwise from root to tip and serve to improve lift like a leading-edge flap. The vortices are particularly strong on ogee-curved wing planforms where the leading-edge sweep angle at the root is acute.

NASA modified the wing of 139208 with plywood ribs and stringers sheeted with aluminum skin to give it a sharp-ogee-curved leading edge. The air inlets were extended forward to accommodate the extended root chord. The area of the modified wing increased 104 square feet over the original F5D wing. The Leading Edge Vortex Phenomena flights were flown by NASA test pilot Fred Drinkwater who gathered the data at the lower end of the Skylancer's speed envelope since the hand-made leading edge was not intended to handle high-speed flight. In his NASA report Drinkwater remarked, "The pilot felt that the directional stability and control characteristics at high angles-of-attack were superior to those of any delta-wing aircraft he had previously flown. The aircraft exhibited no tendency to develop abrupt changes in stability and control."

The lateral and directional control characteristics were so improved over the original F5D wing that Drinkwater was able to lower the approach speed in a carrier-type landing by 10 knots. He did note that the

NASA 212 flying over the San Jose area on final approach to NAS Moffett Field. Vortex emanating from the leading edge is clearly visible as it extends back from the wing root. NASA test pilot Fred Drinkwater discovered that lift generated from the vortex was sufficient to permit a 10-knot reduction in the aircraft's approach speed. Data acquired from this program proved to be valuable in the design of supersonic transports in the mid-1960s. (NASA Ames)

NASA 212 in flight with the ogee wing. The wing was tufted with strips of yarn to study the airflow in flight at various airspeeds and angles-of-attack. Not only was the wing modified, but also the engine air inlets were moved forward to accommodate the leading-edge extension. (NASA Ames)

In 1963 NASA 212 was returned to Ames Research Center. It was redesignated NASA 708, the number it would wear until it was retired in 1968. The yarn tufts used to study airflow can be seen taped to the upper surface of the wing. (Nick Williams Collection)

A close-up of the canopy and windshield of NASA 708. The sharp V-shaped windshield reduced drag by a significant amount and was one of the reasons that the F5D-1 was 63 percent faster than the F4D-1 with the same engine. (Bud Southworth)

To accommodate the extended leading edge of the Ogee wing the engine air inlets were moved forward. This photograph shows the sharp leading edge that was necessary to generate the lift-producing vortex. (Bud Southworth)

Various views of the modified ogee curve leading edge on NASA 708. The leading-edge modification was fabricated from plywood, aluminum, and fiberglass. The new leading edge was not structurally capable of withstanding high-speed flight. This Skylancer was flown at the Ames Research Center and data gathered from the program was very useful in predicting the low-speed flight characteristics of the Concorde supersonic transport. (Bud Southworth)

Late in its career NASA 213, which had been used in the Dyna-Soar project, was redesignated NASA 802. It was used as a chase plane in the Lifting Body Program. Both Dyna-Soar and lifting bodies were precursors to the Space Shuttle. (NASA Dryden Flight Research Center Photo Collection)

A close-up of NASA 708 shows the sharp-ogee-curve leading edge and the blunt engine air inlets, which have been relocated in line with the rear of the canopy. Originally the inlets were located substantially aft of the canopy. (Nick Williams Collection)

NASA 802 flies chase during a Northrop HL-10 Lifting Body launch from its NB-52 Mothership on 30 September 1969. (NASA Dryden Flight Research Center Photo Collection)

NASA 802 undergoes maintenance on 13 October 1967. By 1970 a lack of parts made it economically infeasible to continue operating this last airworthy F5D-1. It was retired from NASA and through the efforts of test pilot Bill Dana it was flown to Dayton, Ohio, and put on display at the Neil Armstrong Museum in Wapakoneta. (Stephen Miller)

ogee wing suffered from some longitudinal instability at the reduced approach speed, which required rapid increases in power. The modified Skylancer became very useful in providing data for the approach and landing characteristics of the Concorde supersonic transport.[14]

The most famous Skylancer research was performed by BuNo 142350 under the Dyna-Soar Program. The Boeing X-20 Dyna-Soar was an early manned spacecraft concept that incorporated a reusable delta-winged vehicle. Boosted into space by a Titan II rocket, it was capable of gliding to a controlled runway landing after re-entering the earth's atmosphere. The X-20 was a forerunner of the Space Shuttle. There were substantial flight-crew safety concerns in the event of a booster malfunction on the pad or shortly after lift-off. This led to a proposal for an auxiliary booster to propel the glider away from the main booster allowing the pilot to gain control and execute a safe landing. However, the hypersonic glider was expected to have a low lift/drag ratio making the landing very difficult. Adding to the problem was the Dyna-Soar's minimally sized windows, which would limit the pilot's visibility.

The Skylancer, having a similar lift-to-drag ratio and wing loading, was expected to approximate the glide profile of the Dyna-Soar. Neil Armstrong, the noted moonwalking astronaut, developed an important launch-abort maneuver for the Dyna-Soar using the Skylancer. Armstrong would begin the abort simulation with a high-speed run at 1,000 feet in clean configuration. At a predetermined point he would pull up into a vertical climb and reduce the power to idle while extending the speed brakes. This simulated the burnout

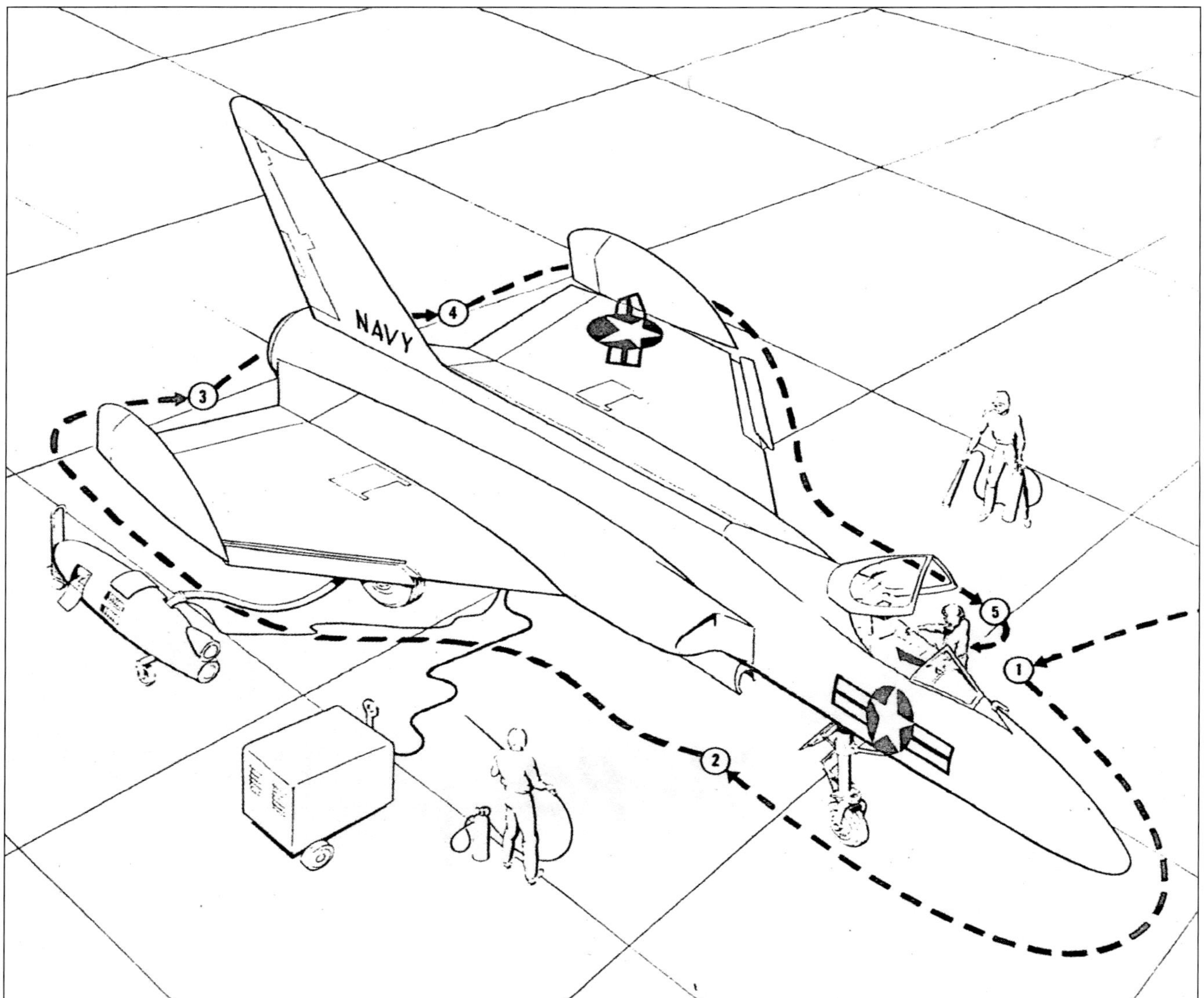

***Illustration from the F5D Flight Handbook** shows the recommended pattern for conducting a preflight inspection of the aircraft. Compare this tech art to the similar illustration for the F4D Skyray shown on page 92. Locations of the engine start unit, ground power cart, and two fire guards are surprisingly similar.* (U.S. Navy)

of the Dyna-Soar's emergency booster rocket. Then Armstrong would pull the Skylancer to an inverted horizontal attitude while extending the landing gear. Finally the Skylancer was rolled erect and accelerated to the proper approach speed for landing.

To simulate the limited visibility of the Dyna-Soar a transparent amber mask was applied to the Skylancer's canopy. Armstrong's flight helmet had a transparent blue visor, which blocked his ability to see through the amber canopy mask. Both 360-degree spiral and straight-in approaches were flown with equal success. Good control was achieved and the visibility limitation did not affect navigation to the landing point nor execution of the flare. Five NASA pilots flew more than 40 simulated launch-abort landings. This provided valuable data for the program but the Dyna-Soar Program was canceled in December 1963 for lack of funds.[15, 16, 17]

BuNo 142350 continued to fly for NASA at Dryden as a chase plane and approach test vehicle for Lifting Body projects. By 1970 parts became critical, and the cost of keeping the Skylancer airworthy exceeded its utility. It was retired and appropriately displayed at the Neil Armstrong Museum in Wapakoneta, Ohio.

Flying the Skylancer

Although the Skylancer never entered operational service, a Utility Flight Handbook and a Temporary Handbook of Maintenance Instructions were published for use during the test flight program. The Utility Flight Handbook covered non-tactical ground and flight operations of the airplane. Since the Skylancer never performed carrier suitability trials only field operations were covered. Many Skylancer procedures were identical to the Skyray, but there were some significant differences.

Skylancer flight operations began with a check of the gross weight and center of gravity for takeoff and the anticipated loading for landing. In the earliest edition of the Flight Handbook (15 January 1956), the aircraft was shown to be limited to a maximum gross takeoff weight of 26,000 pounds and a maximum landing weight of 18,906 pounds. The final edition of the Flight Handbook (1 July 1957) increased the takeoff gross to 27,700 pounds and the landing weight to 26,250 pounds. The center of gravity envelope had a forward limit of 21 percent of the mean Aaerodynamic chord (MAC) and an aft limit of 27 percent MAC.

A preflight inspection commenced at the cockpit entrance on the left side of the fuselage and progressed around the exterior of the airplane. Duct covers were removed, landing gear down-lock pins were removed, and the entire airframe was then inspected for integrity and lack of obvious damage. The Skylancer required a careful examination of the fuselage undersurface around the rocket-bay doors. It was important that the doors closed flush and that the actuating cylinders held the proper pressure. The leading-edge wing slats were checked for freedom of movement, and the landing gear wells were inspected for signs of hydraulic leaks.

The aircraft was entered via a boarding ladder mounted on the left side. A cockpit check of switch positions and warning lights was conducted after the pilot entered the cockpit. Since all four Skylancers carried spin and drag chutes, it was imperative that the chute controls were secured. The Skylancer used the same Pratt & Whitney J57-P-8 turbojet as the Skyray; therefore, engine start procedures were similar. The engine was

NASA 212 at Ames being prepared for mission before it had received the ogee wing modification. Note the extremely large lower mid-fuselage access panel shown open, and markings inside the main landing gear doors. (NASA Ames)

Beautiful side profile view of NASA 212 gives a good indication of the aircraft's nose-high stance on the ground. Unlike Vought's F8U Crusader with its movable variable-incidence wing, the Skylancer's delta planform required a relatively high angle-of-attack to achieve efficient aerodynamic performance when flying at low airspeeds, such as during takeoff and landing. Note how trailing edge of the orange nose section is carefully aligned with the angle of the aircraft's windshield and canopy frame. (NASA Ames)

Nicely proving the commonly held tenet that no two airplane color schemes are exactly the same (sometimes even for the very same airplane), this photograph shows NASA 212 on the ramp at NASA's Dryden Flight Research Center located at Edwards Air Force Base. White placard visible immediately forward of the pilot's boarding ladder reads "AMES" in large block letters. Long red arrow seen below the open cockpit canopy is for readily identifying the emergency canopy release to rescue crews in the event of a crash. (NASA Ames)

Skylancer 213 was also used to test the ball nose cone of the X-15. This nose cone replaced the standard test boom of the X-15 and was essential to the measurement of air data on high-speed re-entry into the atmosphere. (NASA via Tony Landis)

Known as a YAPS (yaw, angle-of-attack, pitch, and sideslip) boom for flight test data acquisition, this device was normally mounted on the nose of various research aircraft. On the F5D-1 modified to test and evaluate the X-15's "Q-ball" nose, the YAPS boom had to be relocated to the aircraft's right wingtip. (NASA via Tony Landis)

spooled up by impingement air from an external cart. When 12- to 16-percent RPM was achieved the throttle was moved outboard to the ignite position and then forward to the idle position. Normally, light-off occurred within 30 seconds and the engine accelerated smoothly to 55- to 65-percent RPM. The tailpipe temperature was monitored to ensure that it did not exceed 630 degreesC, and the oil pressure was expected to reach 30 to 50 psi. If these parameters were achieved the engine was self- sustaining and the external power cart could be disconnected. Since the J57 consumed 2.5 gallons (16.2 pounds) of fuel per minute, taxiing was begun as soon as possible. Advancing the throttle to an intermediate position initiated the roll, and idle thrust was sufficient to keep the Skylancer moving. Steering was accomplished with differential braking.

At the active runway the takeoff roll was commenced in either military or afterburner power. Since rudder control was ineffective below 50 knots, differential braking was used during the early portion of the takeoff roll. The Skylancer had a nose-high ground stance that made rotation of the nose wheel from the runway unnecessary. Pilots were warned that lifting the nose wheel before takeoff would increase drag and greatly lengthen the run. The landing gear was retracted as soon as the Skylancer became airborne to avoid accelerating beyond 250 knots before the 6-second retraction cycle was complete. Failure to do this would exceed the hydraulic limit of the system. The climb was initiated at 380 knots and a climb Mach number of 0.90 above 10,000 feet was considered optimum. Above 10,000 feet the yaw damper, pitch damper, and transonic trim compensator were engaged to ensure high-speed stability. Supersonic flight was achieved easily in the Skylancer. There was far less trim change than that experienced in the Skyray, but a transonic trim compensator was still used to overcome the tuck-under tendency between Mach 0.95 and 1.05.

Maneuvering flight was normal for the F5D but extremely high roll rates could be experienced at high altitude. Full lateral-control deflection could produce a roll of 550 degrees per second. The Skylancer was cleared to perform wingovers, chandelles, Immelmanns, loops, inverted flight, and aileron rolls. The flight controls were powered by a dual hydraulic system. In the event one system failed the second would provide sufficient power for full maneuverability at subsonic speeds. Noticeable airframe buffeting preceded stalls, which started about 40 knots above the stall. As the airspeed diminished a yawing and/or wing heaviness became apparent about 25 knots above the stall. At lower airspeeds the yawing tendency could increase and a spin could develop.

The Skylancer test program was terminated before spin testing was completed; therefore, intentional spinning of the aircraft was prohibited. Douglas test pilot Bob Rahn predicted that the Skylancer would have better spin characteristics than the Skyray, but the Flight Handbook cautioned that spins were expected to be oscillatory in pitch and that the Skylancer may show a tendency to reverse spin direction or enter an inverted spin during recovery. Slow-speed flight above the stall regime was considered very stable with good handling characteristics; however, the pilot was cautioned to avoid high rates of sink, which could be difficult to arrest with military power.

Leading-edge slats automatically extended at airspeeds below 200 knots providing improved lateral and directional stability. Hydraulically operated speed brakes could be deployed at any airspeed but they would blow back above 400 knots. The Flight Handbook recommended that speed brakes be stowed before final approach because of trim changes at slow speeds. Any type of descent that was consistent with weather conditions, airfield position, fuel remaining, or established penetration procedures, was permissible. The transonic trim compensator was turned off at speeds below Mach 0.70 and the yaw and pitch dampers were turned off at altitudes below 10,000 feet. A maximum range descent could be made at idle RPM with the aircraft trimmed to maintain a 250-knot glide. In the landing break the speed brakes could be deployed to reduce altitude or airspeed. Landing gear was extended on the downwind leg. The turn from base onto final was made at 140 knots, and the touchdown was made at 125 to 130 knots.[18]

Different view of NASA 213 fitted with a unique nose-cone adapter for the "Q-ball" nose. This device was essential in measuring airflow data during the X-15's high-speed re-entry into the Earth's atmosphere. (NASA via Tony Landis)

Skylancer NASA 213 was used to fly launch abort profiles for the Boeing X-20 Dyna-Soar manned spacecraft. Here we see NASA ground crewmen awaiting the order to "pull chocks" after the aircraft has undergone final checks on the "last chance" ramp for Edwards' Runway 04. Two large hangars visible in the background are located on the west ramp of Edwards Main Base, but were moved from their original location at South Base in November 1955. (NASA via Tony Landis)

Fitted with a differently shaped device attached to its conical nose-cone adapter, NASA 213 is recovered by a retrieval crew after making an emergency landing on Rogers Dry Lake in apparent winter conditions. Although temperatures can reach 120 degreesF in the summer, winter weather has oftentimes produced snowfalls in the high desert. From the looks of the lakebed, the normally-concrete-hard surface has turned to slush. Note air intake covers in place. (NASA via Tony Landis)

Low wintertime sun angle reveals good detail of the F5D-1's main landing gear structure. This differs from that of the F4D Skyray in overall size and length, and is also mounted in the Skylancer's wing. By comparison, the Skyray's main landing gear is mounted in the wing fillet. Additionally, the Skylancer's main gear door is a one-piece structure, while the Skyray's is a two-piece configuration. (NASA via Tony Landis)

Retrieval crew secures the F5D-1 to its ground transporter, which has apparently backed-up and scooped the airplane onto its carrying trailer. Note size of the transporter's rugged trailer tires that look well suited to the task at hand. (NASA via Tony Landis)

NASA personnel pose with NASA 213 in the main hangar at the Dryden Flight Research Facility. Famed NASA test pilot Bill Dana is seen standing third from left. Dana, the last man to fly the X-15, was hired on the very day that NACA (National Advisory Committee on Aeronautics) became NASA (National Air and Space Administration). Note the F5D's large square photo-calibration marking aft of cockpit. (NASA via Tony Landis)

Another view of the recovery effort for NASA 213 on Rogers Dry Lake. Aircraft is now securely mounted on the recovery transporter, and is ready for the slow ride back to the NASA ramp. (NASA via Tony Landis)

Convair's Second-Generation Delta

From the outset the F-102A was considered an interim solution to the Air Force high-performance interceptor requirement. It was to be replaced by a second-generation Ultimate Interceptor in 1954. But extensive developmental problems delayed the F-102A's entry into operational service until April 1956. That placed the Air Force and Convair under enormous pressure to accelerate the development of the Ultimate Interceptor, originally designated the F-102B. In concept, the -B was to be an F-102A with an improved engine and electronics package. However, the aerodynamic problems with the F-102A were so profound that a complete airframe redesign was essential.

The redesign included lengthening the nose and moving the cockpit forward, repositioning the air intakes, and changing the fuselage contours to take better advantage of Whitcomb's Area Rule principles. Compounding the problem was the Wright Aeronautical J67 engine that was expected to power the F-102B. A license-built variation of the British Bristol Olympus, it proved to be so troublesome that the Air Force abandoned it and switched to the Pratt & Whitney J75.

An early F-106A (then called a YF-106A) on the ramp at Edwards Air Force Base during flight test. This airplane was more than 17 feet longer and was powered by an engine capable of producing 8,500 more lbf than the F5D-1. (NASA Dryden Flight Research Center Photo Collection)

Nevertheless, 17 F-102Bs were ordered into production in November 1955.

The Air Force still subscribed to the Cook-Craigie philosophy of procurement to expedite new aircraft development. Under Cook-Craigie no experimental prototypes were built, and low-volume production was started at the outset of the contract. Testing was performed on early production aircraft, and modifications were retrofitted to aircraft on the production line. This proved to be a costly mistake with the F-102 when nearly $40 million of production tooling had to be scrapped because of design changes.

The F-102B contract called for two aircraft to be delivered by the end of 1956, and the remaining 15 by mid 1957. By June 1956 it was apparent that the Ultimate Interceptor was becoming far more complex than a re-engined F-102A. Recognizing that the Ultimate Interceptor was evolving into an entirely new airplane, the F-102B designation was changed to F-106 Delta Dart. The contract specified that the F-106 was to have all-weather capability, a combat ceiling of 70,000 feet, a combat radius of 375 nautical miles, and a top speed of Mach 2.0. The combat ceiling was later reduced to 55,000 feet but the combat radius was increased to 633 nautical miles. In August 1956, five months before the first F-106 flew, the Air Force expanded the order to include a two-place trainer variant, the F-106B. Unlike the side-by-side TF-102A, the F-106B was designed with tandem seating to ensure high-speed performance.

This front view of a test F-106A clearly shows the size and location of the enlarged inlet ducts that were a major feature in the airframe redesign. The progressively cambered leading edge is also apparent in this view. (NASA Dryden Flight Research Center Photo Collection)

The first F-106A (tail number 56-451) flew at Edwards AFB on 26 December 1956 with Convair test pilot Richard Johnson at the controls. The second aircraft joined the test flight program on 26 February 1957. The program moved rapidly with the first Air Force flight taking place on 29 April 1957 when the aircraft was flown to 57,000 feet at a speed of Mach 1.9. By May 1957 the Air Force approved conditional acceptance of the F-106As in the flight test program and began Category II testing under the direction of Capt. Iven Kincheloe. However, Category II tests revealed several performance deficiencies. Acceleration from Mach 1.0 to Mach 1.7 took nearly 4-1/2 minutes, and another 2-1/2 minutes to Mach 1.8 while 2,000 pounds of fuel was consumed. The aircraft was considered not tactically usable at speeds greater than Mach 1.7. The test aircraft also experienced fuel starvation under certain conditions and the ejection seat was considered inadequate at supersonic speeds. Most of 1958 was spent on engineering fixes for these problems. Category II testing did not conclude until June 1959 when the F-106 was placed in operational

In these top views the pure delta-wing configuration can be seen. The F5D-1 retained the F4D-1 planform of rounded tips with a swept trailing edge. The F-106A wing provided 661.5 square feet of area compared to 557 square feet for the F5D-1. In this view the area-ruled fuselage is clearly evident. (NASA Dryden Flight Research Center Photo Collection)

service with the 498th Fighter Interceptor Squadron—nearly five years late.

Only 277 Delta Darts were built, fewer than any other Century Series aircraft. In spite of its developmental difficulties the F-106 proved to be an exceptional airplane. On 15 December 1959 Maj. Joe Rogers set a world absolute speed record of 1,525.695 mph in an F-106A. This record stands as the fastest speed ever achieved by a single-engine aircraft. The Delta Dart remained in operational service for 29 years and, like the F-102A, it had a "life after life" as a drone and NASA research aircraft.[19]

The Convair F-106A was much larger than the Douglas F5D. It was nearly 71 feet long and had a gross weight of more than 34,500 pounds. Its Pratt & Whitney J75-P-17 engine produced 24,500 lbf in afterburner and propelled the F-106A to a maximum speed of 1,328 mph. The F5D was only 53 feet 6 inches long with a gross weight 27,700 pounds. It was powered with the same J57-P-8 engine that was used in the Skyray rated at 16,000 lbf in afterburner, which provided the F5D with a maximum speed of 1,094 mph. The Skylancer never received the J57-P-14 engine that was planned for production aircraft, which was expected to provide 17,200 lbf in afterburner giving the F5D a significant increase in speed.

A side view of the two-place F-106B that was used for training purposes. Unlike the side-by-side configuration of the two-place F-102, the F-106B used a tandem configuration. As a result the performance of the F-106B was nearly the same as the single place F-106A. (San Diego Air & Space Museum)

The last operational unit to operate the F-106 was the 119th Fighter Interceptor Unit of the New Jersey Air National Guard. Its last F-106A was retired in August 1988 giving the Delta Dart an operational life of nearly 30 years. (San Diego Air & Space Museum)

EPILOGUE

ASSESSMENT OF THE SKYRAY

Somewhere around the world, seven days a week, 365 days a year, a Douglas F4D-1 Skyray was spotting on an aircraft carrier catapult, as seen here. This scene took place thousands of times during the career of the U.S. Navy's first dedicated carrier-based jet interceptor. This chapter gives a historic retrospective of the aircraft's successes and failures for the record, and defines its rightful place in aviation history. (Nick Williams Collection)

Few aircraft have sparked as much debate about their merits as the F4D-1 Skyray. Many respected naval aviators, historians, and authors revered this aircraft, yet others reviled it. Here is an examination of the criticism and praise that has been written, a review of the Skyray safety record, and reflection on the Skyray's place in history with the advantage of hindsight.

The Criticism

CAPT Gerald O'Rourke, USN (Ret.), a combat veteran of World War II and Korea, Commanding Officer of the USS *Independence* (CVA-62), and test pilot, was one of the Skyray's most outspoken critics. Summarizing his overall assessment of the Skyray, he stated, "I once had respect and admiration for the Ford. After years of retrospect, however, I feel that it was a pretty bad airplane, which should have been bypassed in favor of either a better F3H or an F5D version. Myself, Tim Wooldridge, John Glenn, Tony Nollet, George Watkins, and a few others were Patuxent River test pilots at a time when we could have forced the issue. We did not, for one reason or another. I feel we were either wrong or lazy on the matter, and we should have demanded a long series of fixes and improvements, which probably would have resulted in scrapping of the Ford in favor of the F5D."[1]

In an amusing article, "The Douglas Navy F4D Ford,"[2] O'Rourke summed up the overall characteristics of the Skyray: "Its primary design goal, when it was first thought up, was for speed, acceleration, and maneuverability. The airplane had no tail, just a single, broad, manta-ray-shaped wing from which a vertical tail protruded. The size of the plane was small, the construction fairly lightweight, and the overall design extremely compact. The manta-ray wing proved to be great for turning and climbing, but it was an anathema to the turbulent shock waves of transonic air.

"Night and all-weather flying means instrument flying for which good inherent flight stability characteristics are ideal. The F4D had almost none. All-weather flying means a lot of headwork on the part of the pilot, a lot of reference to papers and charts, and radios and lights and switches. The F4D cockpit was jammed and crammed. Switches were haphazardly placed. The huge radarscope hid behind the stick. The instruments were grossly misplaced. Light reflections danced around the multi-faceted windscreen and canopy like a go-go dance pavilion.

"The autopilot, which could have been a godsend to the harried pilot, never worked worth a damn. The airplane was just as happy flying sidewise as straight ahead. The afterburner lit with a bang that curdled your buttocks, and the control stick was short, stubby, hesitant, halting, heavy, and jerky in motion, and obscured the radarscope anyway.

"Originally planned for a relatively low-powered engine, the Ford's designers felt that the thick wing would more or less ooze through the sonic wall, or, failing that, it wouldn't fly that fast very often to be a big problem. When a different, more powerful engine was substituted in final design, the effect was devastating! The plane could, would, and often did get into this speed regime, where it reacted like a sledgehammer against a sheetmetal building. It would fly supersonically, and not too uncomfortably, once you got it through the transonic region, but you had quite a ride—and steeply downhill—to get there. There was much consternation in test circles about how a fix could be made. A whole book could be written about a wizard electronic Trim Change Compensator, which was supposed to provide the aerodynamic grease to slide through to supersonics. Like so many other brave ideas for the airplane, it never worked....

"Even worse, the F4D carried a relatively miniscule amount of fuel, but was equipped with an engine capable of emptying a tank car in minutes. The acceleration and rate-of-climb were phenomenally fast—but waiting just ahead was the near-impenetrable sonic barrier, where the Ford would buck, tuck, shake, quake, hump, and bump as it tried to batter its way through on brute force alone. Fearsome enough high in a sun-filled sky, the experience at night, in rain on instruments close to a stormy ocean, is best described as sheer unrelenting terror!"

O'Rourke then described the shortcomings of the Aero 1A Weapons System, which relied on vacuum tube electronics to aim and fire unguided air-to-air ballistic rockets with a noticeable lack of precision. However, he conceded that things got better with the arrival of the Sidewinder: "Luckily for the Ford, the Sidewinder air-to-air, heat-seeking missile came into operational use at about this same time. With the Sidewinder, a computer wasn't essential, as long as the target's hot tailpipe exhaust could be 'seen' by the infrared missile head. So, in actual fleet use, Ford pilots would use both the good radar and their own eyeballs to get into a Sidewinder launch position on a target's tail. In daylight, or on good clear nights, these tactics would probably have worked fairly well for actual combat, but if there was any rain or clouds, or if the target speed was in the transonic region, the Ford was just another toothless tiger."

O'Rourke saved his most caustic criticism for an indictment of the Skyray's landing characteristics: "Then there were carrier landings in the good old Ford. Always interesting, any one could be a memorable, breathtaking experience. A safe, slow landing speed had not been high on the initial design priorities for the bird. As a result, the first Fords were pretty speedy on landing. By the time the radar, the computer, and the drop tanks had added their weight, the Ford's landing speed had escalated like the cost of living today. When the real, but intangible pilot pucker factor for night and rain and a pitching deck were also figured in, the Ford was coming aboard at speeds close to the 140 knot region, which is even today considered close to the speed of light by experienced carrier landing signal officers.

"At any speed below about 170 knots, the Ford was one of the worst Dutch Rollers ever to fly. The reason for this was simple, easily understood nowadays. The plane was short, or close coupled, with no separate tail surface. The wings were very large for the size of the plane, very fat by modern standards, and altogether too efficient at producing lift at slow speeds. The vertical tail was too small, and tended to get blanked out by the wings at the high angles-of-attack required for slow-speed flight. As a result, in each approach the slightest disturbance induced by rough air or a rough pilot would make the Ford swing from side to side. As it did, the advancing wing increased its lift, the opposite one decreased its lift, and the plane would start to roll. The pilot would push rudder, which wouldn't help much, but then put in some correcting elevon... This would both slow the up wing and turn the plane back toward straight ahead, usually just about in phase with a naturally slow tendency of the plane to correct itself. This would induce an overshoot, usually augmented by the sticky, heavy, jerky stick movement, and away he'd go—right into a series of wallowing, half-roll, half-sideslip maneuvers that made the bird look sort of drunken."

Finally, O'Rourke listed Skyray attributes that he labeled several other little nasties. The first was the control system that was hydraulically driven, but had a manual backup system of cables and pulleys that idled along all the time. O'Rourke blamed the jerkiness of the controls and poor harmony on the drag from the manual system.

Second, he pointed out that at the end of its carrier career, Skyrays had a tendency to shed their tail hooks during traps—a very messy situation at best. Also, portions of afterburner nozzle had a tendency to stick shut, producing a wild ride from the vectored thrust produced by the asymmetric nozzle. Rapid throttle movements caused an imbalance between the engine compressor and the air inlets. This created a succession of *chug, bang!* noises that could be very disconcerting, but harmless. The airstream at the wing automatically actuated the leading-edge slats. Unfortunately, one wing could experience higher velocity airflow than the other. Thus the slats didn't always deploy in unison. In a sideslip or a Dutch Roll the slats would slap in and out exaggerating the uncoordinated flight condition.

The worst feature, according to O'Rourke, was the outboard location of the guns. Any off-axis disturbance would be magnified at the gun location. In O'Rourke's words, "As a result, very few pilots could hit the side of a barn with the damn things."

Jerry O'Rourke was not the only Skyray critic. He was joined by noted test pilot Phil Oestreicher who flew F4Ds both in the Fleet with VMF(AW)-114 and at the Naval Test Pilot School. "I had, at the time I was there in 1966, more time in the Skyray than all the instructors and other students put together. Actually, I think I was the only student in my class [#44] who had flown it. TPS made good use of its assets and thus I became the unofficial 'F-6A [post 1962 designation for the F4D-1] Chief Instructor Pilot.' There was no such title bestowed. I did not make up any portion of the curriculum and there was nothing in it specific to the Skyray. Mainly, I flew as instructor on familiarization flights [generally two flights per aircraft type new to the student.] I used whatever suitable chase plane—in which I was current—that was available such as another Skyray, an F-8B, or an A-4B."[3]

On 10 August 1966 Oestreicher, as part of his test pilot training, conducted an exercise on the maneuvering stability of the Skyray using a hand-held force gauge and cloth measuring tape to determine the stick forces during steady turns, sudden pull-ups, and steady pull-ups while flying at 300 knots IAS at 30,000 feet. The purpose of the exercise was to evaluate the service suitability and specification conformance of the F-6A with regard to maneuvering stability. On 12 September Oestreicher continued the exercise investigating the transonic flying qualities of the Skyray. His report stated, "The observed changes of longitudinal stick force during accelerations through the transonic region were excessive and very distracting to the pilot. The tactical value of the F-6A airplane in the transonic region is seriously compromised because the pilot is required to devote a major portion of his attention to basic air work at the expense of other duties such as fire control system operation."[4]

Addressing the question of why the Naval Test Pilot School employed its fleet of Skyrays for so long, Oestreicher made some interesting observations: "My opinion was that it was kept as an interesting and not particularly dangerous 'horrible example' or a 'Don't ever buy another one like this' airplane. Structurally, the Skyray was something of a paradox: a very strong landing gear attached to a thin skinned, fatigue-prone, and (might as well say it) flimsy airframe. In VMF(AW)-114 we were forever stop-drilling skin cracks, putting on doublers, etc. Somewhere I saw a Skyray picture, supposedly at high speed—which showed virtually every skin panel bulged out (or in) from air loads, a far cry from the F9F series!

"It was a very poor handling airplane in the transonic regime but, to me at least, handled reasonably in the landing pattern so long as one was willing to learn how to fly it properly. Despite the turn coordination feature of the upper (or servo) rudder, the Skyray's high adverse yaw, high dihedral effect, and low directional stability combined to keep it out of the typical 'feet on the floor' jet category. Only the Aeronca Defender and Champion (in my experience) required more pilot rudder input to turn properly at pattern speeds! A turn downwind immediately after takeoff (as in a field carrier landing practice operation) was best initiated with a small amount of rudder, which was followed by lateral stick. This kept the ball reasonably centered throughout the turn. I never flew a 'Cutlass' so I don't know how that tailless fighter handled, but all this rudder pushing was not necessary in the F9F, FJ, F3D, TV, and F8U or, for that matter, AD-5 airplanes.

The Skyray handled very well on the final portion (on the mirror) of a carrier or runway landing. Sink-rate control was very precise and easy, and the airplane held the trimmed angle-of-attack (and, hence, the airspeed) very well. When I cruised with the airplane in 1959, we did not have angle-of-attack indications in the cockpit. These were incorporated by the time I flew the airplane at TPS. Even discounting my additional pilot experience by that time, it was obvious that believing the AOA indicator and ignoring the buffet made the approach turn much more comfortable and enhanced the precision of control down the glide scope to touchdown."[5]

Several years later Oestreicher restated his concerns about the Skyray: ". . . the Skyray was a decent but not spectacular accomplishment for the time [early 50s] given the then-existing general knowledge of aerodynamics. However, there are some things about it that seem inexcusable. Why in the world would someone put such a small vertical tail with so much sweep on a maneuvering airplane?... Sure, we had learned more in the interim like area ruling, etc., but adequate directional stability was hardly a new concept. The bizarre spin behavior early in the test program should have given someone a clue that all was not well. It is probably fitting that the last Ford crash involved a spin."[6]

George Watkins, an NATC test pilot who was involved in the Skyray's Service Acceptance Trials in 1957–1958, expressed the following misgivings: "I think it's important to understand a couple of different things about an engineering test pilot. He has a responsibility to the Naval Air Systems Command (and to the Navy) to ensure that any new model airplane not only meets the contract specifications and guarantees submitted by the airframe contractor, but also to ensure that the fleet pilot is getting a damn good flying machine that performs its assigned fleet mission in a superb manner.

"The test pilot has an equal responsibility, as a given model project pilot, to report all deficiencies found in stability and control, handling characteristics, performance, carrier suitability, and serviceability of that new model airplane before it goes to the fleet as an acceptable production model. The F4D-1 had very poor transonic trim change characteristics, right in the Mach range where the airplane would be expected to engage in air-to-air combat. It also had very poor lateral/directional/roll characteristics, among other things. This is one of the main reasons why I was against production and even further testing of the F5D-1 model. This airplane, in December 1956, exhibited the same poor flying qualities, as did the Skyray.

I wasn't really an F3H fan, but it did have good flying qualities and handling characteristics, and was a good, stable platform in the carrier approach. I enjoyed flying the F4D very much, but like the F8F-1 Bearcat before it, it just didn't provide the fleet with a good stable-platform, high-performance fighter in the air-to-air role. And back in 1955 when we were flying the XF8U-1 at Edwards and getting level-flight Mach numbers of 1.5 and higher with essentially the same engine as the F4D, it was a little difficult to get too pumped up over the Skyray's subsonic performance limitations and capabilities."[7]

The Praise

The Skyray's most avid proponent was, of course, its designer, Ed Heinemann. When he started work on the Skyray in the late 1940s, Heinemann was already well respected in the aviation industry, but future successes made him an aeronautical legend. By all accounts Heinemann regarded the Skyray as his crowning achievement. Most of his work up to that time involved the design and production of largely conventional airplanes. He was noted for very durable, highly reliable machines built in large numbers that provided competent performance and exceptional longevity. Examples that come to mind are: the SBD Dauntless, A-20 Havoc, A-26 Invader, and AD Skyraider.

In the late 1940s Heinemann had the good fortune to obtain a valuable and challenging contract from the Office of Naval Research for high-speed research aircraft. Although this was not a large-volume production commitment, it did allow Heinemann and his team to focus on the upper limits of performance. The outcome of this work was the D-558-1 Skystreak, which broke the world speed record in 1947, and the D-558-2 Skyrocket, which achieved Mach 1.88 in 1951 and became the first aircraft to fly at Mach 2.0 in 1953. The Skystreak was particularly valuable to the Navy since this airplane put them on a par with the Air Force in transonic and supersonic knowledge. The early Skyray work was motivated by the success of these high-speed research aircraft.

In his biography Heinemann recalled an incident that illustrates the admiration naval aviators held for the Skyray: "At the Association of Naval Aviation convention in San Diego in 1977, a former *Forrestal* flyer cornered me. He had never flown the F4D Skyray and, in fact, had been a Skyraider pilot in the early 1960s aboard CVA-59.

"'I loved the Spad,' he told me, 'but when I checked aboard and saw an air wing firepower demonstration for the first time, I developed a bad case of envy. I really wished I was flying Fords.'

"'We were in the Med,' he went on, 'and had some dignitaries on board. To start the show the Skyrays were loaded onto all four catapults, and the pilots lit their afterburners simultaneously. It was ear shattering. The whole ship seemed to vibrate. You could feel the heat rise up from the blast deflectors.

"'In quick intervals each Ford was shot into the sky and seconds later the shapes of the airplanes vanished in the distance, leaving behind their exhaust trails. These converged toward each other like a quartet of comets, burning hot, and climbing damn near straight up until they disappeared altogether. It was exhilarating. After that, the other events were anticlimactic.'

"'That's what they were designed for,' I said proudly. 'To get to altitude in a hurry. You know,' I went on, 'right here at North Island the Skyray really made a name for itself. Of course, that was over 20 years ago.'

"I told the officer about All Weather Squadron Three, the only Navy unit assigned to...NORAD, the North American Air Defense Command. CDR Gene Valencia, the Navy's third leading ace in World War II, with 23 aerial victories to his credit, was in that outfit and helped lead it to two consecutive top honor awards in 1957 and 1958, as the best interceptor squadron in NORAD.

"The aircraft was indeed a success. Revolutionary in its design, it was a real performer. I was humbled and honored in 1953 when President Eisenhower awarded me the Collier Trophy, presented annually for 'the greatest achievement in aviation in America.' The Skyray was that 'greatest achievement.'"[8]

In 1982 Heinemann, who had achieved elder statesman status, was invited by the editors of *Air International* to give his thoughts on ideal fighter design. The result was an insightful article, "A Veteran Designer's Thoughts on Fighters." Reflecting on his success with the F4D he wrote, "By December 1958, however, when the 419th and last production F4D was to be rolled out, the Skyray was acknowledged as a thoroughly competent interceptor with excellent high-altitude dogfighting capabilities and outstanding climb performance. Its low-speed characteristics were perhaps unusual, but its pilots soon accustomed themselves to these."[9]

Bob Rahn, the former fighter pilot who had been appointed chief engineering test pilot for the XF4D-1 program, acquired far more experience with early Skyrays than anyone. In his biography Rahn stated his first impressions of the radical jet: "My first flight in the aircraft was in October 1951, after Thaw (Russ Thaw) had made 25 flights. After observing Thaw's flights, especially his landing approaches, I wasn't so sure that I wanted to check out in this airplane, as this type of aircraft, with its modified delta-wing planform, has a very high angle-of-attack, which looks as if the airplane is about to stall at any moment. After my checkout, however, this high angle-of-attack did not bother me due to the excellent visibility over the nose. I soon realized that this aircraft was just what I had been looking for in a fighter since my flying days in the Spitfire. The F4D was a fighter pilot's dream as it combined: 1) maneuverability at altitude, 2) extremely high rate-of-climb, 3) short takeoff and landing distances, 4) excellent low-speed handling characteristics, 5) extremely high service ceiling, and 6) high-level flight speed."[10]

Bill Gunston, a Royal Air Force instructor pilot in the late 1940s, became the technical editor of *Flight International*, an authoritative British aviation periodical. In that capacity he visited nearly 300 aircraft establishments of all types in much of the world, and wrote more than 50 books and countless articles on the aerospace industry. He devoted a chapter to the Skyray in his book, *Early Supersonic Fighters of the West*, which begins: "I have included it in this book for several reasons. One is that it was a unique and extremely effective answer to a very difficult problem. Another is that it makes an instructive and entertaining story. Perhaps the best reason is that I think it is fair to describe the Skyray as the first supersonic fighter in the world. One could argue this indefinitely, but the design of the Skyray began as early as that of any other supersonic fighter, it flew much earlier than any other supersonic fighter (though with a temporary engine of inadequate thrust for exceeding Mach 1), and when given a more powerful engine it did exceed Mach 1 on the level (though the F-100 and MiG-19 had done so previously). And it was also a successful programme, because more than 400 joined the U.S. Navy and Marine Corps."[11]

Describing the significance of the Skyray, Gunston wrote, "During World War 2 the ignorance of aerodynamicists on what shape aircraft ought to be for flight at transonic Mach numbers was profound." There had been virtually no work on this issue in the United States or Britain; however, Germany, as was discovered at the end of the war, had worked frantically on the problem. Heinemann and his team at Douglas were encouraged by the German data, but they found several dead ends in the tailless delta-wing designs of Alexander Lippisch. The thick wing that was required to house the engine and pilot made supersonic flight impossible.

"The breakthrough," wrote Gunston, "was the decision to blend the wing into the fuselage with a huge tapering fillet of increased chord. This concept gave the best tunnel drag measurements and also overcame the problem of finding space for adequate fuel tankage. (After the Skyray, it was apparently forgotten until the Air Force FX programme of 1966 made several companies rediscover it, and today it is seen in the two most important Air Force supersonic aircraft, the B-1 and the F-15, as well as light fighters YF-16 and YF-17.)"

When Heinemann's work matured to the point where it looked like it might produce a practical fighter, the Navy drafted a specification that Gunston considered unique. "It was, of course, to be capable of reaching 40,000 feet within five minutes of taking off, and...intercepting and destroying a 500-mph bomber. This specification was, I believe, the first formulated by any Navy in response to the threat posed by the jet bomber carrying nuclear weapons. Thus the Douglas design can be considered the first naval interceptor—as distinct from fighter—in history."

Gunston noted that the Skyray development contract was awarded in 1948 at a time when BuAer decided to fund several radical fighter programs. The tailless Chance Vought Cutlass, the delta-wing Convair SeaDart, the swing-wing Grumman Jaguar, and the tail-sitting VTOL Lockheed XFV-1 and Convair XFY-1 were part of the array of new programs. "Out of the whole lot, though the Cutlass did later see limited service, only the Skyray can be considered a success."[12]

The Skyray Safety Record

In July 1986 a report was issued by KLD Associates, which contained an analysis of F4D-1 Skyray safety data.[13] The data was accumulated by the Naval Aviation Safety Center and covered 455 mishaps over the 9½-year operational life of the Skyray. These incidents ranged from minor accidents that resulted in no injury or aircraft damage to major events that produced fatalities and total loss of aircraft. Of course, landings produced the most mishaps, 36 percent, while takeoffs produced 15 percent, in-flight mishaps accounted for 16 percent, wave-offs accounted for 5 percent, taxiing accounted for 2 percent, and mishaps not incident to flight accounted for 26 percent of the total. Surprisingly, the majority of landing mishaps were field landings, a total of 95, while carrier landings produced only 67.

The report focused on the 250 mishaps that occurred during takeoffs, landings, or wave-offs. It was concluded that the takeoff mishaps (both field and carrier) showed no design problem with the F4D. Pilot error, carrier-related equipment failures, and engine problems were the dominant cause of takeoff incidents. Likewise, wave-off mishaps were related, for the most part, to pilot judgment and not aircraft configuration. However, landing accidents (both field and carrier) were found to be design related.

The report stated, "By far the most numerous mishaps are found in disembarked or field landings... Almost half of the reported mishaps were classified as Arrest/Barrier Engagement and involved essentially no damage to the aircraft. The data show that the majority of mishaps were caused by wet runways, poor judgment on the part of the pilot, and engine problems. Rain was a major degrading factor, not only because of its deleterious effect on braking capability but also because of poor forward visibility through the F4D windshield. This problem was noted several times and must be classified as an airplane design problem. Except for the visibility problem, airplane design characteristics do not appear to have been a significant factor in field landings mishaps."

Carrier landings were quite another story. Carrier-landing mishaps were accompanied by substantially more damage than field landing mishaps, which is in keeping with overall Navy experience, but it seems to have been more severe for the F4D than other Navy aircraft of the period. The report concluded that the F4D's high carrier-landing accident rate suggests that the Skyray was more difficult to bring aboard than other aircraft of its generation. The low speed lateral-directional characteristics compounded by the visibility issues at night and in heavy rain were clear design deficiencies.

In a separate study it was determined that 95 Skyrays were lost or experienced ejections during its operational life (essentially the same 9½-year period as the KLD study). This represents 22 percent of the entire F4D-1 fleet. The F-102A experienced a loss of 247 aircraft over its operational life, or approximately 25 percent of its fleet. Meaningful comparisons are difficult since the F-102A had a longer service life, it saw combat action, which accounted for some losses, two foreign governments with different training and maintenance standards operated it, and it was not subject to carrier operations. Nevertheless, it is apparent that both first generation tailless deltas experienced a high attrition rate. However, neither was as dangerous as the conventionally configured F-100 or F-104. In their first 750,000 hours of operational service the North American F-100 suffered 287 total losses, and the Lockheed F-104 lost 160 examples. By comparison the Convair F-102A lost only 107 aircraft in the same period of operation. Since the Skyray experienced fewer losses than the F-102A over its total operational life, it can be argued that the safest airplane of the group was the Douglas F4D.[14]

Reflections

In the 60 years since the Navy ordered the XF4D-1s, the Skyray has been regarded as an aeronautical masterpiece. It was a winner of several world records, and yet has been considered as a mistake with numerous flaws and safety defects. In truth, the Skyray's value lies somewhere in between. The Skyray was designed to perform a mission that was considered impossible when the specification was issued. It was required to climb to 40,000 feet in just five minutes, then maneuver supersonically and destroy a bomber flying at high speed in the next five minutes.

The airplane that Douglas built was indeed able to accomplish that. However, to achieve that level of extreme performance, several characteristics were designed into the airplane that proved to be troublesome in its future service. First, the airplane needed to be as light as possible, thus the airframe was skinned with very thin-gauge aluminum. Over time the skin became a maintenance nightmare and Heinemann himself regarded the skin thickness as a mistake.

Second, the airplane needed a large wing area to meet its climb and maneuverability criteria. A tailless delta configuration was chosen to maximize the wing area within the most compact shape. This unique configuration was the source of control and stability issues throughout the Skyray's life.

Third, the airplane was required to fly above Mach 1, but little was known about supersonic flight at the time the Skyray was designed. In fact, the XF4D-1 mock-up was approved less than 18 months after Yeager's first supersonic flight. None of the aerodynamic refinements such as Area Rule, ultra-thin wings, low-drag canopy shapes, or thin intake ducts were employed in the Skyray.

And fourth, the Skyray was designed as a pure interceptor. It was intended to climb fast, maneuver at altitude, and have limited range. Unfortunately, the Skyray was rarely used in that role. It was forced into the role of all-weather general-purpose fighter, a role for which it was not well suited. It was even used as a ground-attack fighter, which caused structural failures on occasion.

The Skyray's shortcomings were recognized very early on, and in 1953 Heinemann launched a complete redesign resulting in the F5D-1 Skylancer. By all accounts the Skylancer was a superb airplane overcoming all of the Skyray's flaws. Gone was the yaw instability, gone was the transonic pitch/trim issue, gone was the compressor stall problem at high altitude, and gone was the maintenance and structural problem with Skyray's thin skin. Furthermore, the Skylancer, using aerodynamic refinements, improved the Skyray's top speed by 63 percent with the same engine. It is reasonable to think that if the Navy ordered the Skylancer into production and the proposed J79 engine was installed, it would have matured into a weapons system superior to the Vought F-8.

Perhaps it is fitting that Jerry O'Rourke, who wrote the most scathing criticism about the Skyray, eloquently stated the Skyray's importance to Naval Aviation: "Nevertheless, in all fairness and with 20 or more years of retrospect to soften the absurdities, the Ford had some very fine characteristics and provided a wealth of experience, which served so well in later years in the Phantom over Vietnam and the Sinai. The bird was rugged and the J57 Pratt & Whitney engine was a real live brute, with a reputation for dependability and durability unparalleled even today. The big manta-ray wing was great in a high altitude dogfight when it could be horsed around and still provide lift at almost any attitude. The same characteristic could be used to make super-tight turns, and the tremendous power available could be used to climb like a rocket—better than any other design, U.S. or foreign—of that era. The APQ-50 radar was one of the first really good modern airborne sets, and far outshone the miserable installation and rotten computer weapons system it was tied to. Its very direct descendants were used in almost all the F-4 Phantom models, where they have been head-and-shoulders superior performers.

"Along with the F3H Demon of those days, the Ford trained the cadre of pilots who were later to gain fame in Phantoms, and who were to make all-weather fighter operation from today's carriers almost a *no-sweat* routine. The costs were high. Many fine young men were lost in F4D accidents over the years. There's no recalling these lost souls. But there are many, many more alive, kicking, and proudly pushing Phantoms and the new F-14 Tomcats through the night skies today who would not be half so professional, so competent, or so confident had it not been for the venerable, incongruous, imponderable Douglas F4D Skyray of the mid 50s."[15]

THE SURVIVORS

There were no F4D Skyrays or F5D Skylancers that remained airworthy after the year 1970. However, there are seven Skyrays and two Skylancers currently on static display throughout the United States. These rare and historic aircraft are shown here.

124587 displayed at the U.S. Naval Armament and Technology Museum, China Lake NAWS, Ridgecrest, California. This is the second XF4D-1 prototype which set the world speed record and was used for testing the General Electric J79 turbojet. (Nick Williams Collection)

134806 displayed at the U.S. Naval Aviation Museum, Pensacola, Florida. This was the last airworthy Skyray. It was attached to the Naval Test Pilot School at NAS Patuxent River, Maryland. (Author's Collection)

134748 displayed at the Pima Air and Space Museum, Tucson, Arizona. This is one of the first twelve production Skyrays. It wears the markings of the Douglas Testing Division aircraft. (Author's Collection)

134764 displayed at Naval Air Test and Evaluation Museum, Lexington Park, Maryland. This was the first Skyray assigned to the Naval Test Pilot School. (Stephen Miller)

134836 displayed at the New England Air Museum, Windsor Locks, Connecticut. Note "USAF" painted on the right wing. (Robert Boswell via Author's Collection)

139177 displayed at the Marine Corps Air and Ground Museum, Quantico, Virginia. This was assigned to the Naval Test Pilot School and was one of the last airworthy Skyrays. (Stephen Miller via Nick Williams Collection)

134936 displayed at the Pueblo Weisbrod Aircraft Museum, Pueblo Colorado. This Skyray was attached to VF-881 at NAS Olathe, Kansas and was the last airworthy reserve F4D-1. (Author's Collection)

142350 displayed at the Neil Armstrong Museum, Wapakoneta, Ohio. This was the last Skylancer built, and the last Skylancer to remain airworthy. It was flown by Neil Armstrong when he was still a test pilot at NASA's Dryden Flight Research Center, and is reportedly the last airplane he flew before becoming a NASA astronaut. (Stephen Miller)

139208 displayed at Ontario Municipal Airport, Ontario, Oregon. This was the Skylancer that received the ogee-curved leading edge to investigate vortex lift. It was flown at NASA's Ames and Dryden facilities by test pilot Fred Drinkwater. (Bud Southworth)

Appendix II

F4D & F5D Color Schemes

The profile and plan view drawings on these pages were produced by aviation artist Jack Morris, who meticulously researched the markings you see depicted here. These drawings show not only the colors for every single prototype, flight test, and operational aircraft, but in many cases, variations of those markings identified by unit deployment and year.

Pertinent graphic images such as test or operational emblems, unique nose art, distinctive one-of-a-kind patch insignia, and unit shields also accompany the profile drawings where appropriate. With such a unique wing shape, markings for the Skyray and Skylancer upper surfaces could not go unnoticed, and are shown here as well.

Looking at these beautiful drawings makes one hearken back to a time in aviation history when military aircraft, and especially U.S. Navy carrier-based aircraft, served as veritable three-dimensional canvases for unit insignia and color scheme designers. Colorful aircraft were not only functional from an inflight visibility standpoint, but also easily stood out on a crowded aircraft carrier deck or air base ramp.

Whether modeler, historian, or enthusiast, we hope you will find these images to be helpful in identifying the many interesting and colorful phases of F4D and F5D history and development.

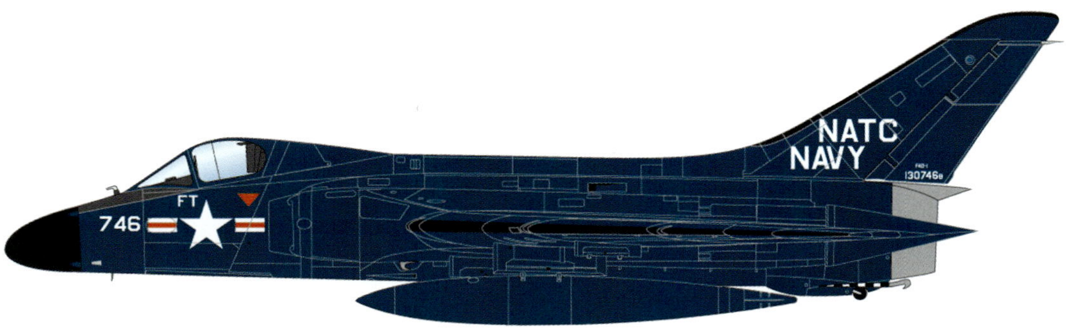

NATC Flight Test, NAS Patuxent River, MD
February, 1956

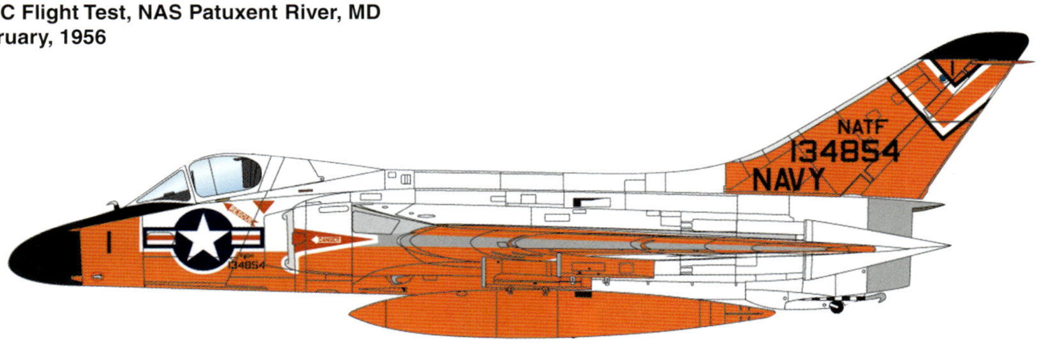

NATF NAS Lakehurst, NJ
May 1961

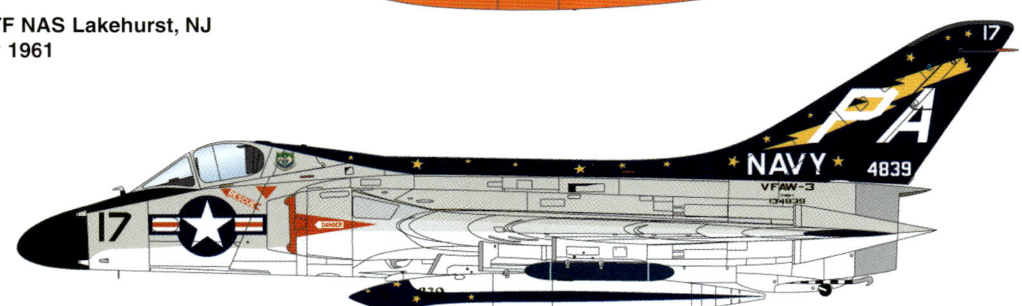

VF(AW)-3 NAS North Island, CA
1960

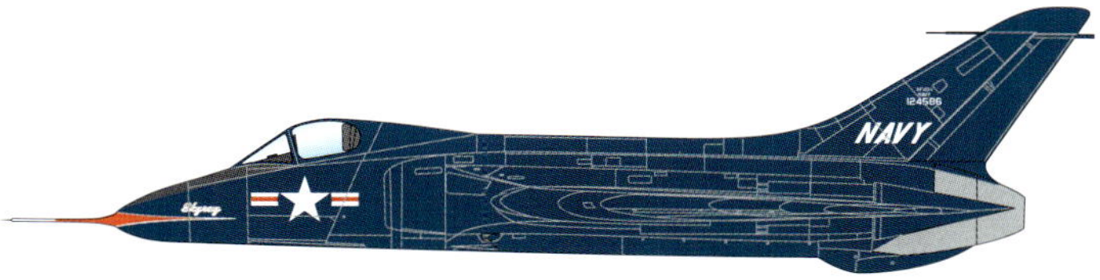

#1 Prototype Douglas Testing Division
Edwards AFB, CA January, 1951

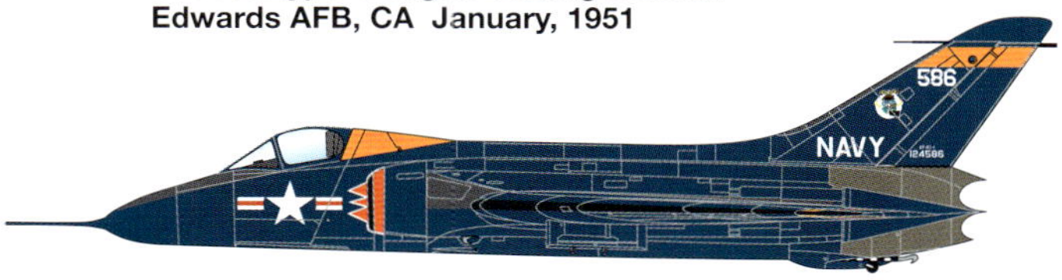

#1 Prototype Douglas Testing Division
NAF China Lake, CA   1955

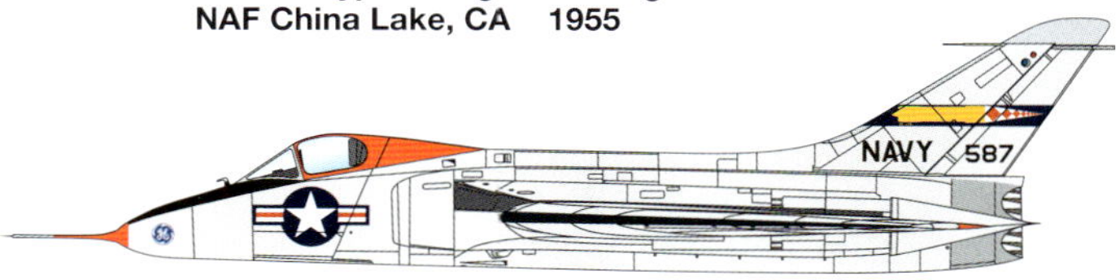

#2 Prototype/GE J79 Engine Test
Edwards AFB, CA December, 1955

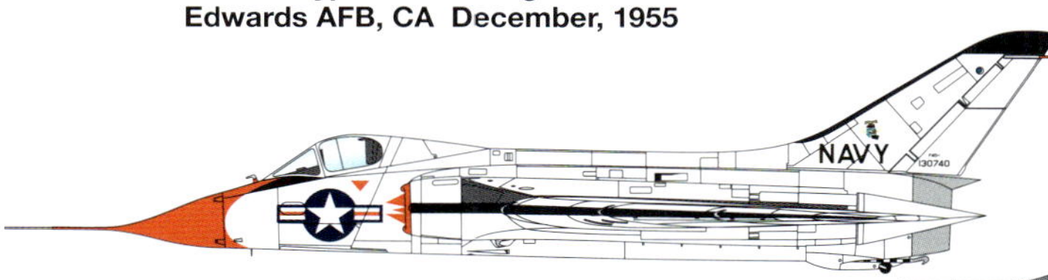

Douglas Testing Division, Edwards AFB, CA
Summer, 1954

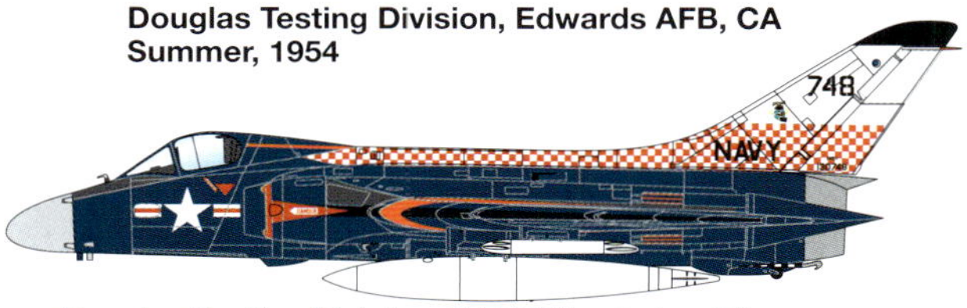

Douglas Testing Division, NAF China Lake, CA
May, 1957

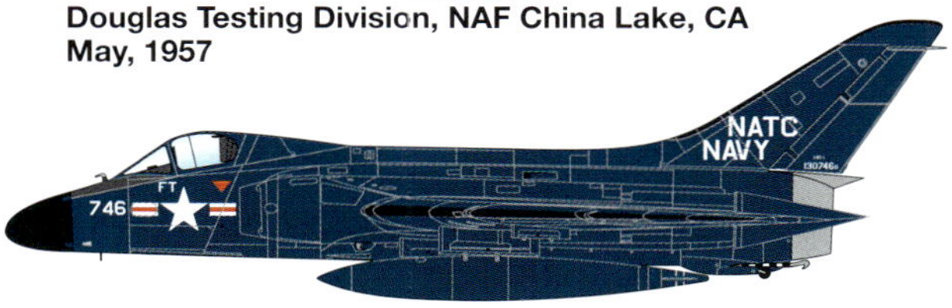

NATC Flight Test, NAS Patuxent River, MD
February, 1956

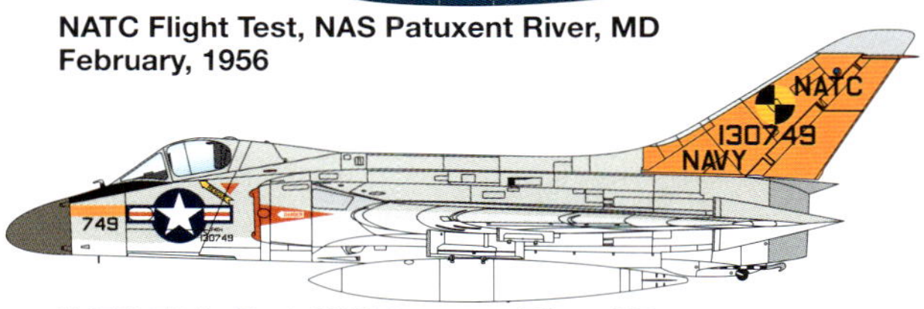

NATC Flight Test, NAS Patuxent River, MD
Autumn, 1960

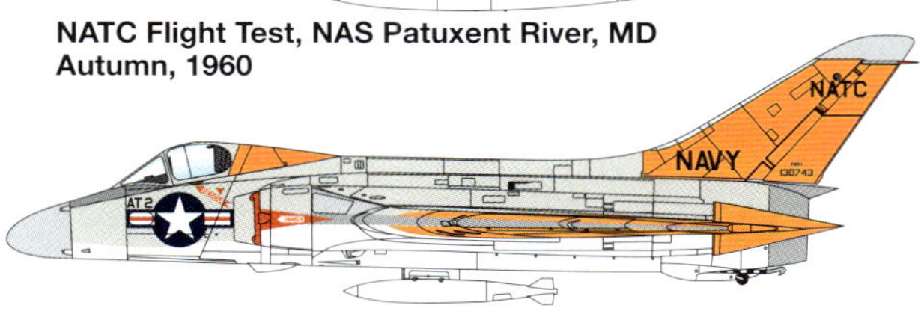

NATC/Armament Test, NAS Patuxent River, MD
1958

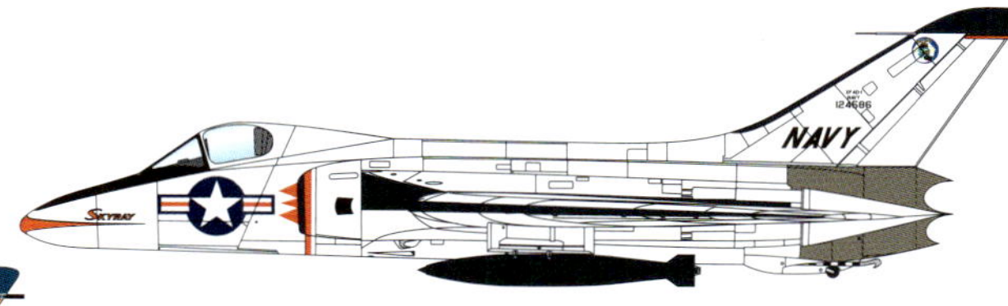

#1 Prototype Douglas Testing Division/NATC
USS Coral Sea (CVA 43) October, 1953

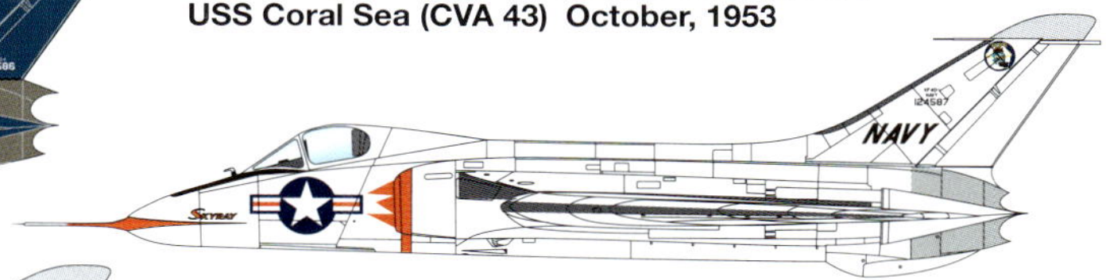

#2 Prototype Douglas Testing Division
Edwards AFB, CA   December, 1953

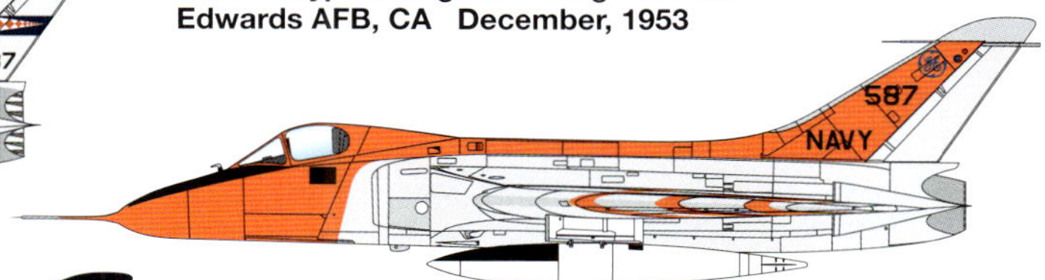

#2 Prototype/GE CJ-805 Engine Test
Edwards AFB, CA April, 1958

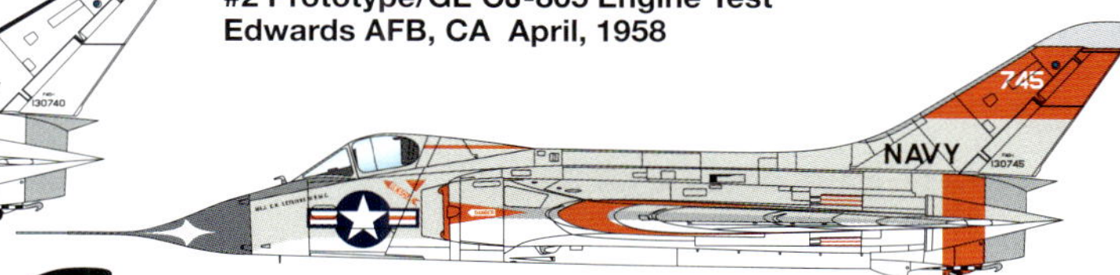

Douglas Testing Division, NAF Point Mugu, CA
May, 1958

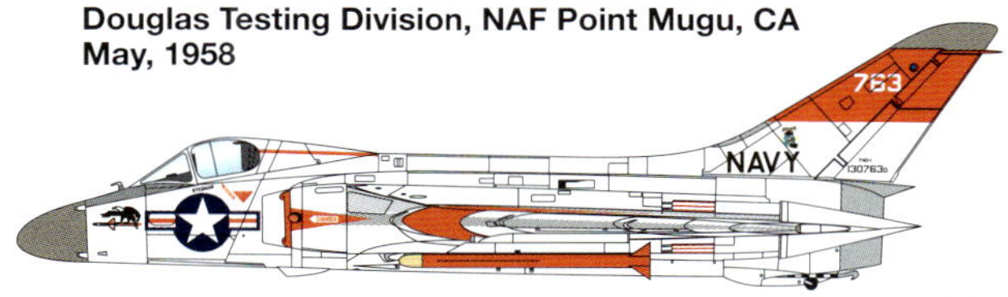

Douglas Testing Division, NAF China Lake, CA
August, 1956

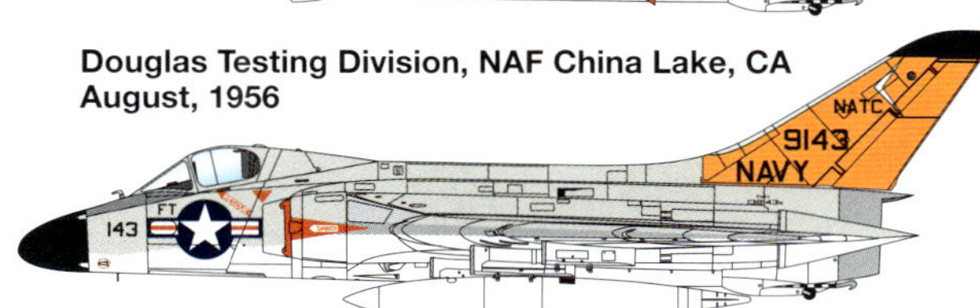

NATC Flight Test, NAS Patuxent River, MD
1959

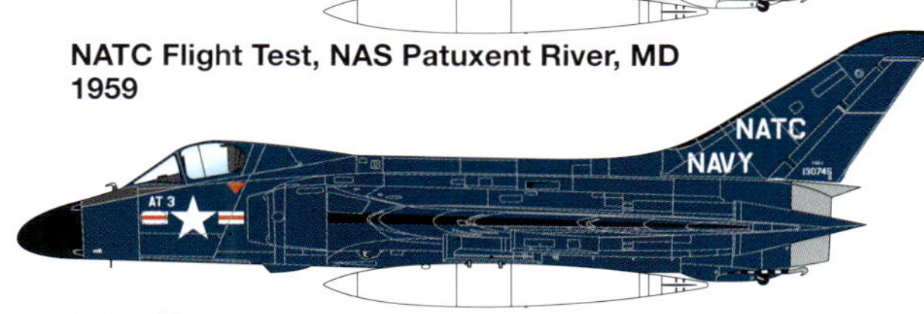

NATC/Armament Test, NAS Patuxent River, MD
April, 1957

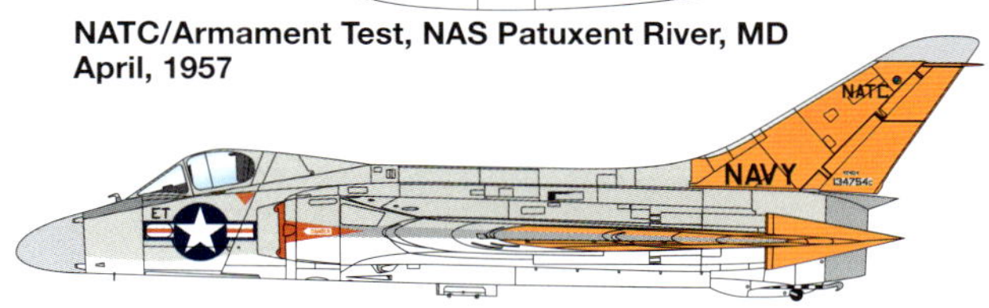

NATC/Electronics Test, NAS Patuxent River, MD
1958

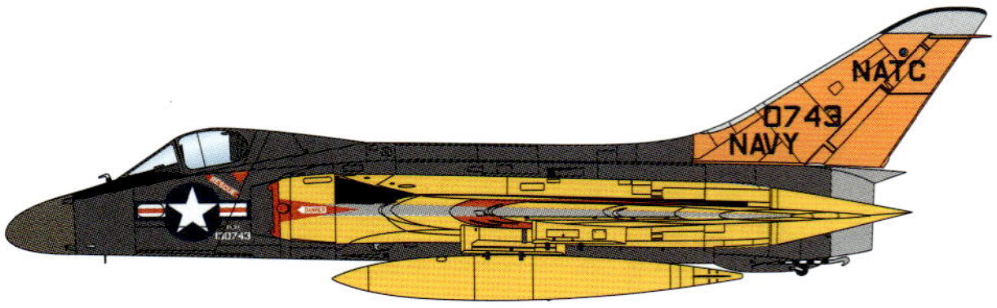

NATC NAS Patuxent River, MD
May, 1960

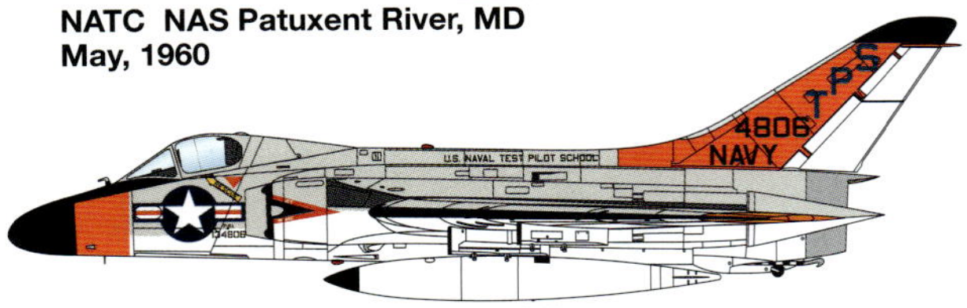

USNTPS NAS Patuxent River, MD
December, 1966

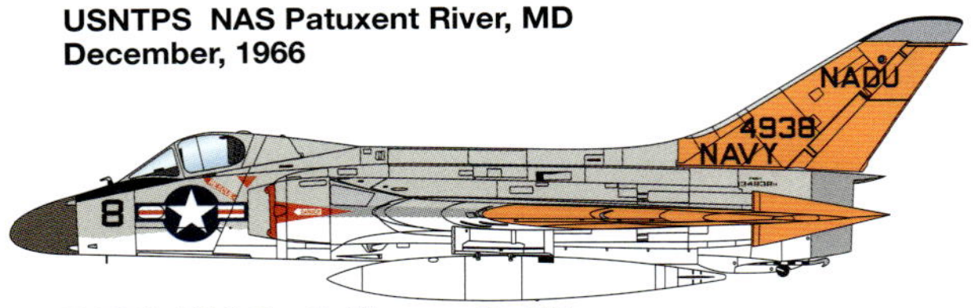

NADU NAS South Weymouth, MA
May, 1957

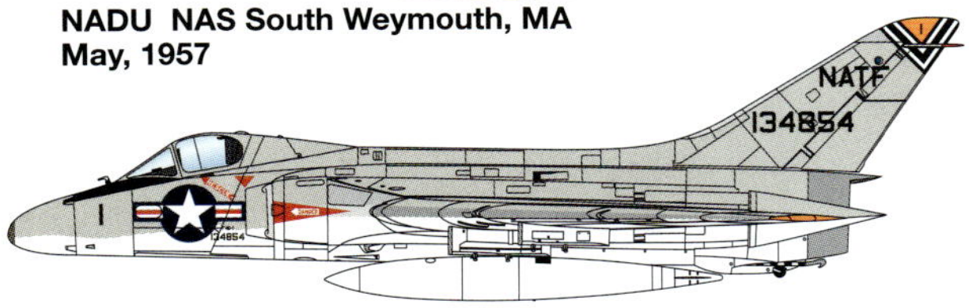

NATF NAS Lakehurst, NJ
March, 1960

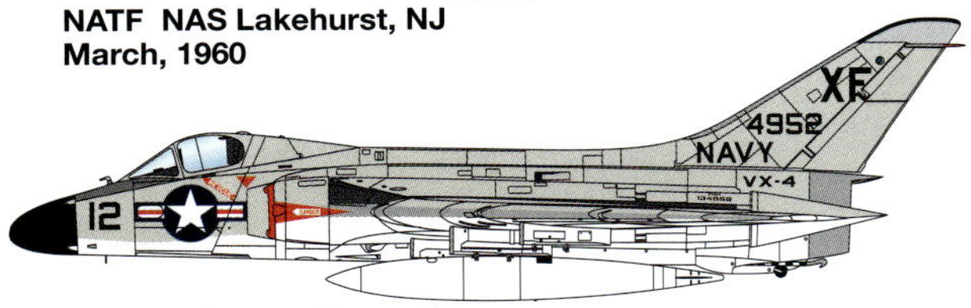

VX-4 NAS Point Mugu, CA
April, 1960

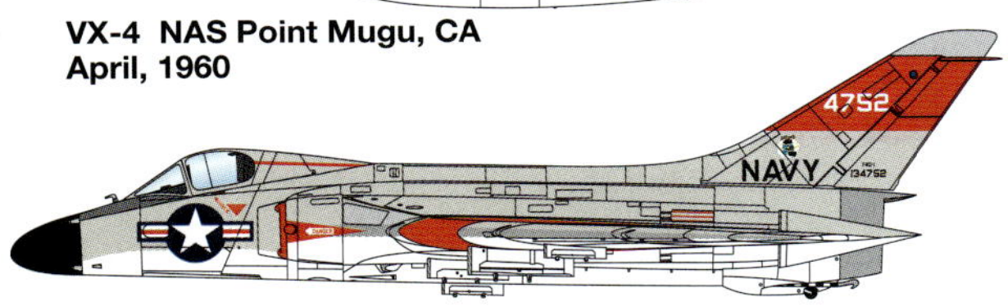

NOTS NAF China Lake, CA
May, 1956

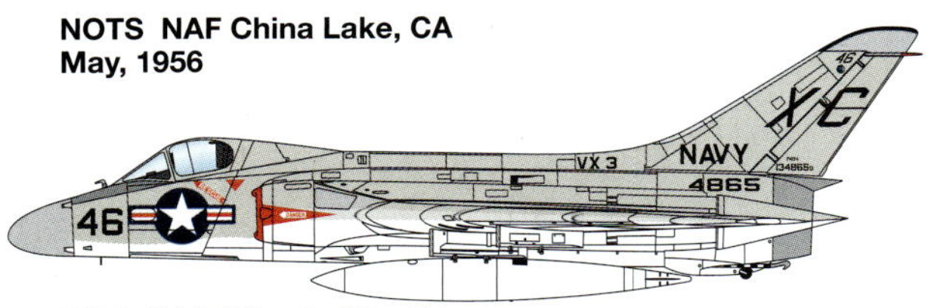

VX-3 NAS Atlantic City, NJ
February, 1957

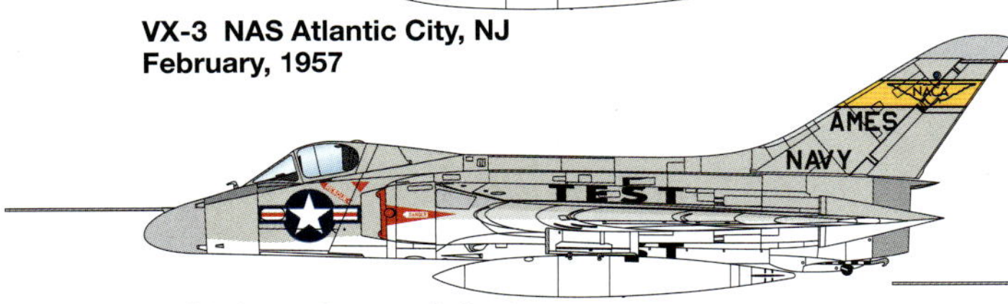

NACA Ames Research Center
NAS Moffett Field, CA 1957

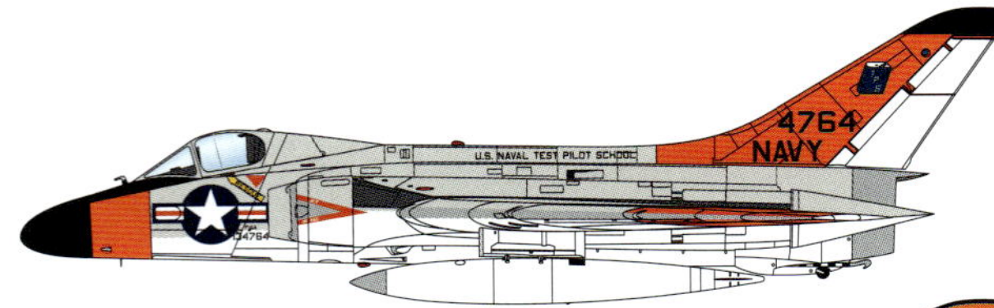

USNTPS NAS Patuxent River, MD
July, 1966

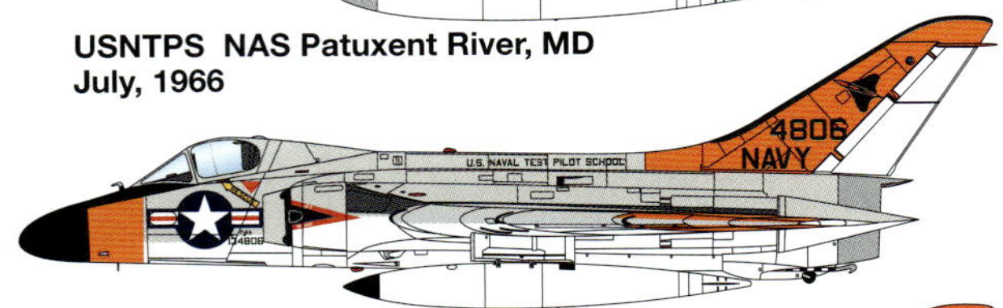

USNTPS NAS Patuxent River, MD
1967

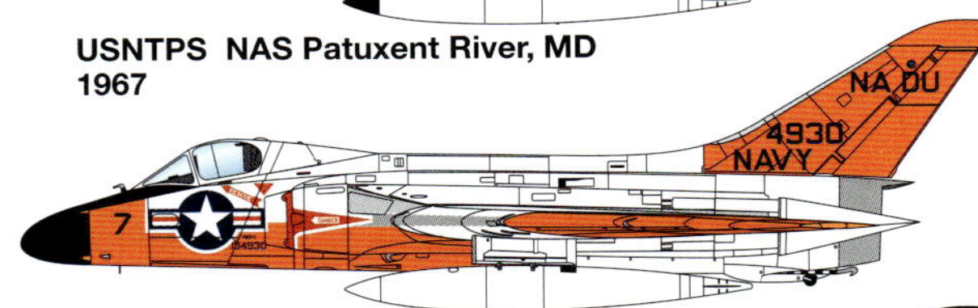

NADU NAS South Weymouth, MA
April, 1960

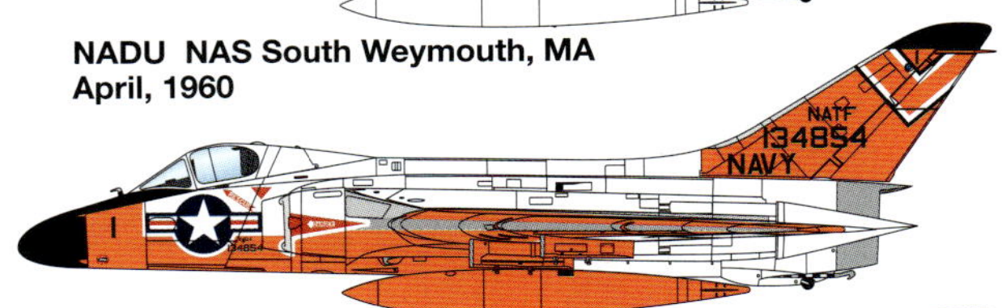

NATF NAS Lakehurst, NJ
May, 1961

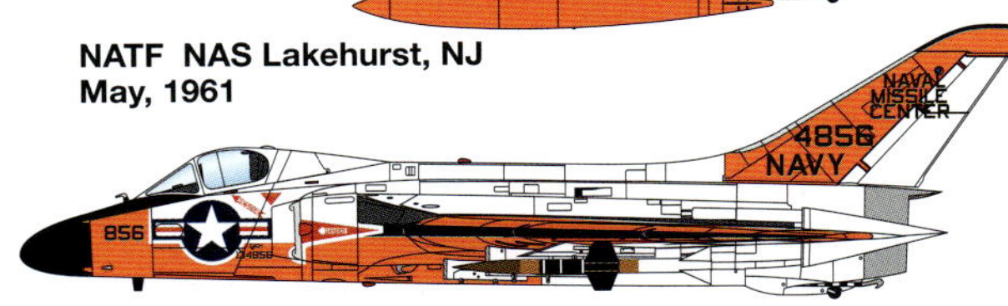

NMC NAS Point Mugu, CA
1961

NOTS NAF China Lake, CA
June, 1957

NOTS NAF China Lake, CA
1959

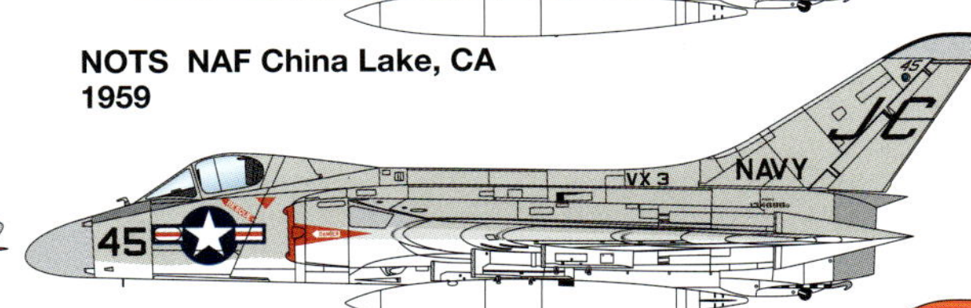

VX-3 NAS Oceana, VA
April, 1958

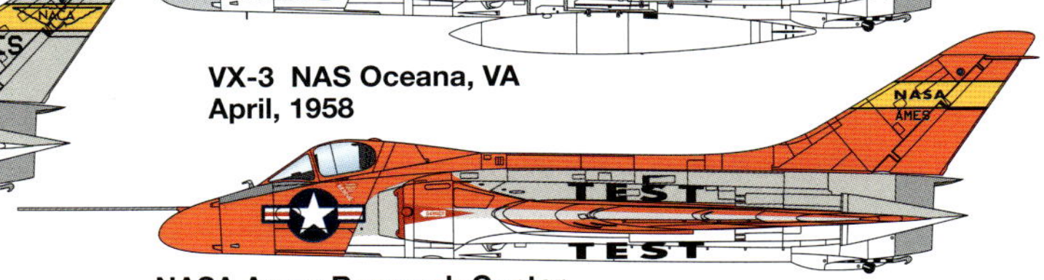

NASA Ames Research Center
NAS Moffett Field, CA May, 1959

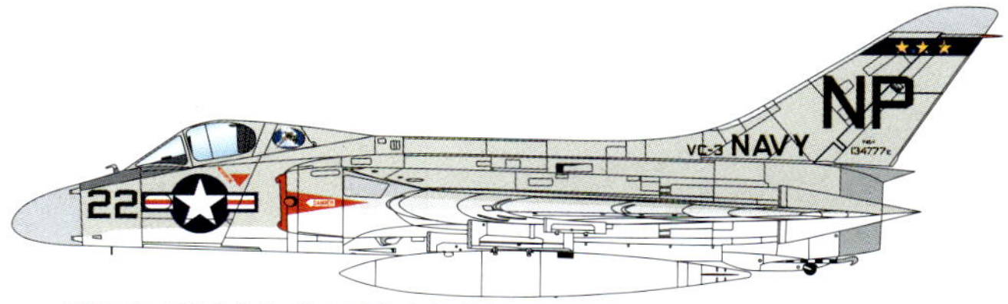

VC-3 NAS Moffett Field, CA
May, 1956

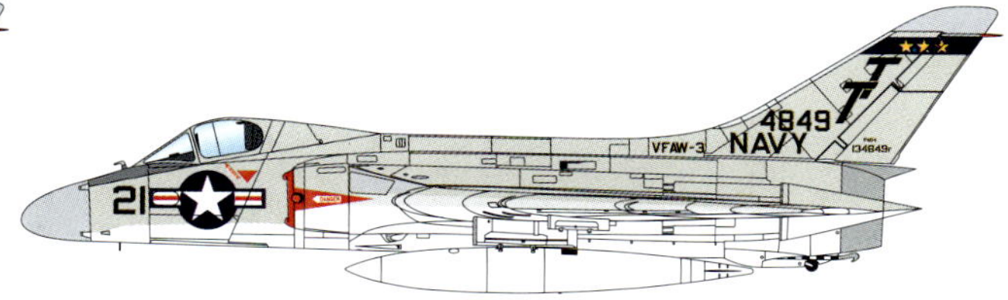

VF(AW)-3 NAS Moffett Field, CA
May, 1957

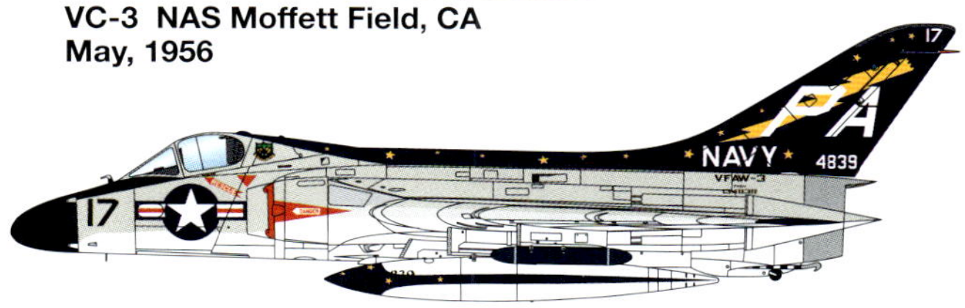

VF(AW)-3 NAS North Island, CA
1960

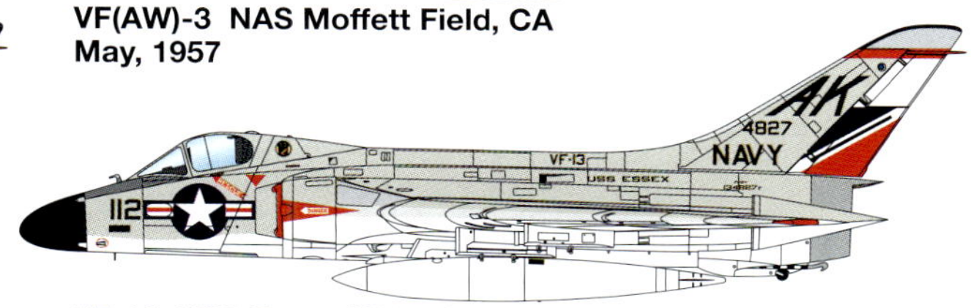

VF-13 USS Essex (CVA 9)
August, 1960

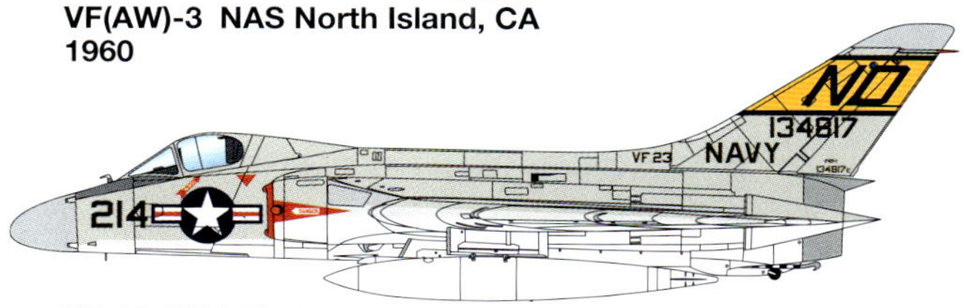

VF-23 NAS Moffett Field, CA
August, 1957

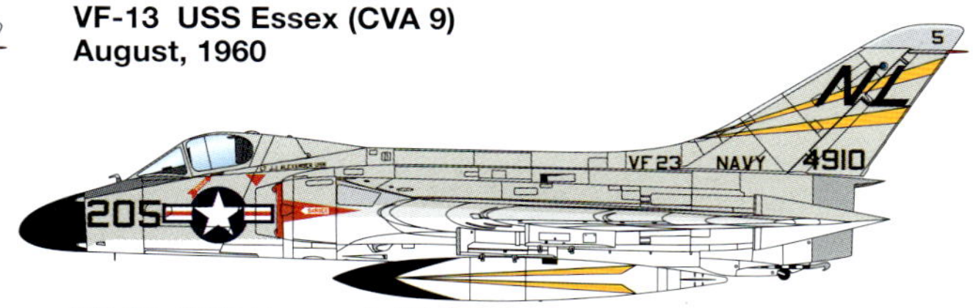

VF-23 USS Hancock (CVA 19)
June, 1958

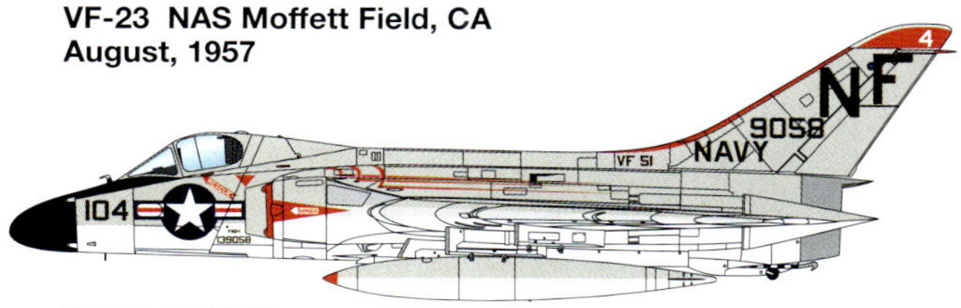

VF-51 NAS Miramar, CA
Summer, 1959

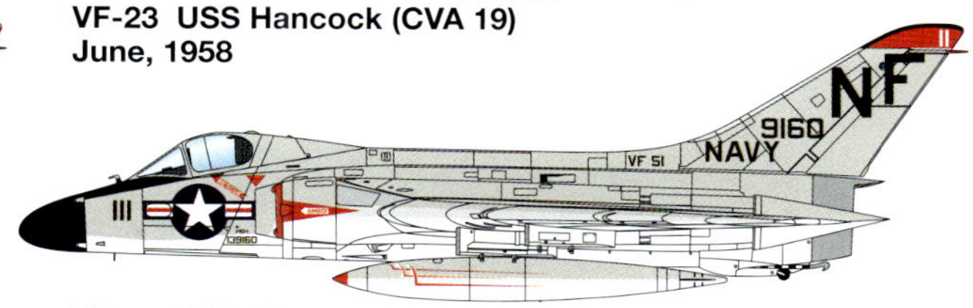

VF-51 USS Ticonderoga (CVA 14)
Spring, 1960

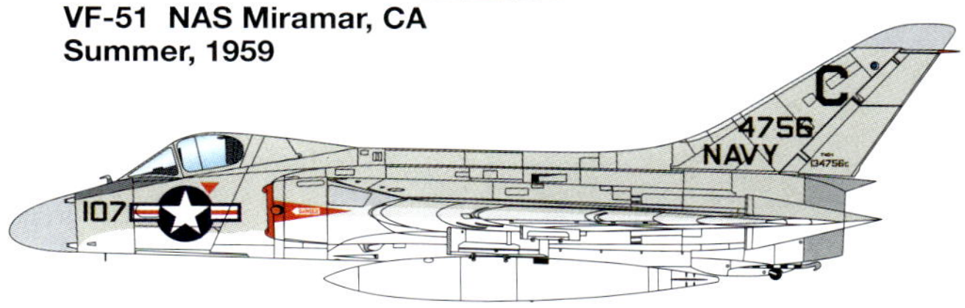

VF-74 NAS Oceana, VA
May, 1956

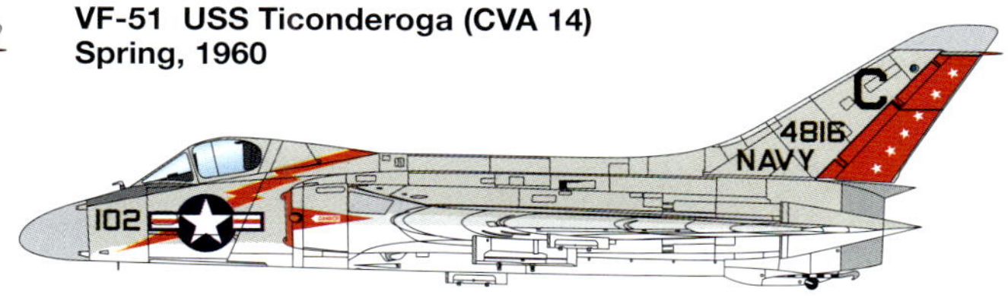

VF-74 NAS Oceana, VA
November, 1956

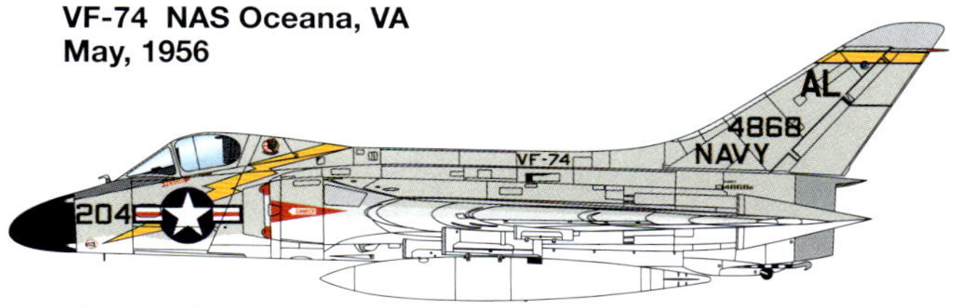

VF-74 USS Franklin D. Roosevelt (CVA 42)
Summer, 1957

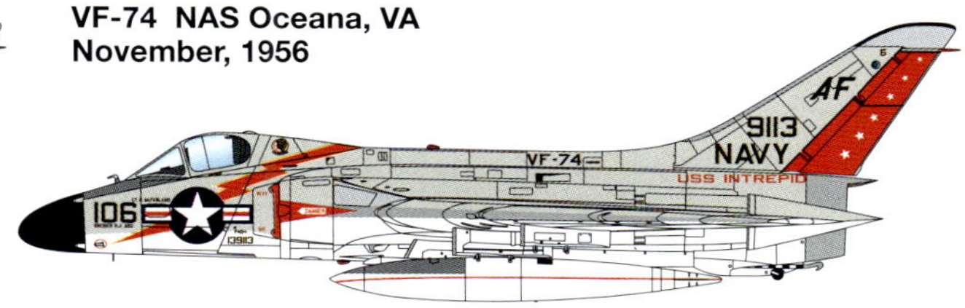

VF-74 USS Intrepid (CVA 11)
April, 1961

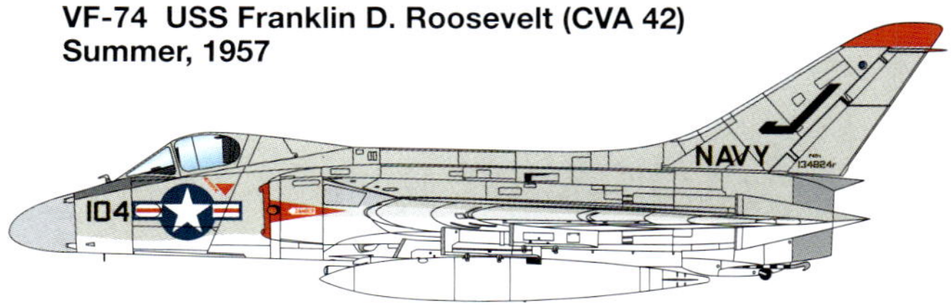

VF-101 NAS Cecil Fld, FL
December, 1956

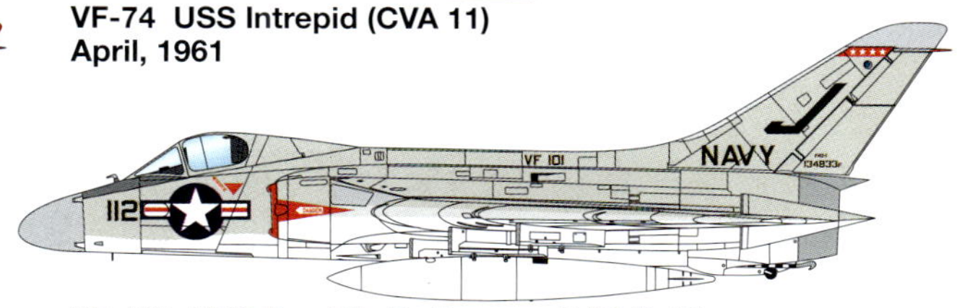

VF-101 USS Franklin D. Roosevelt (CVA 42)
May, 1957

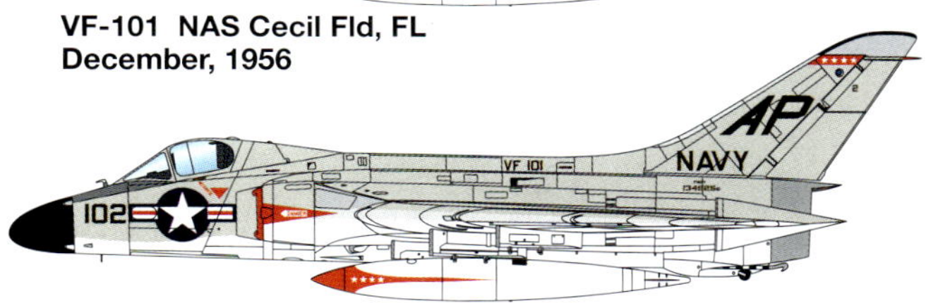

VF-101 USS Saratoga (CVA 60)
1957

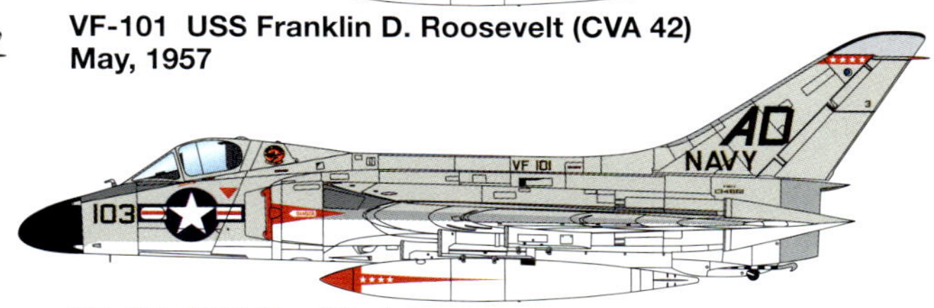

VF-101 NAS Key West, FL
1962

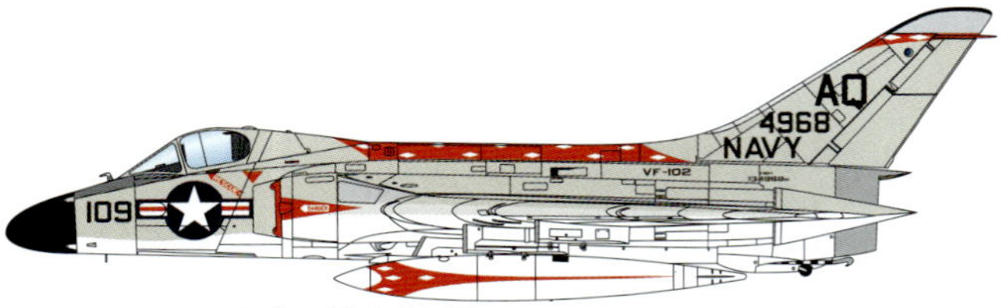

VF-102 NAS Cecil Field, FL
December, 1957

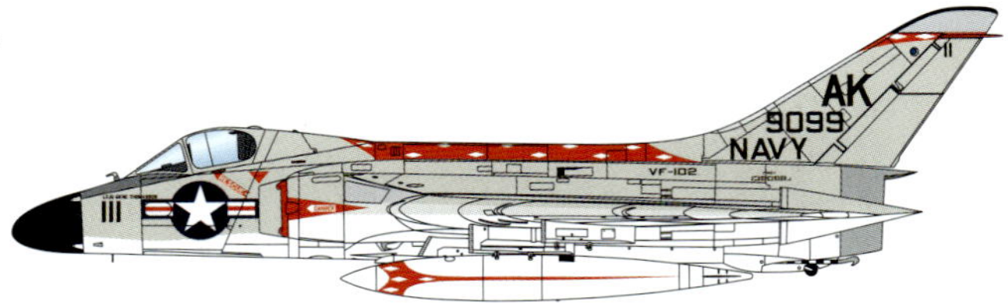

VF-102 USS Forrestal (CVA 59)
November, 1958

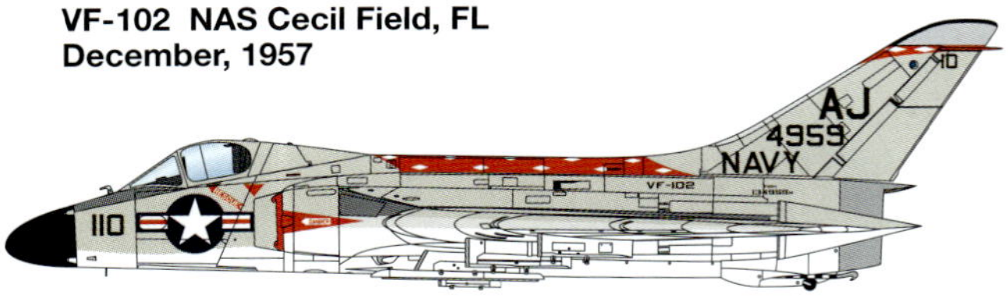

VF-102 NAS Cecil Field, FL
December, 1959

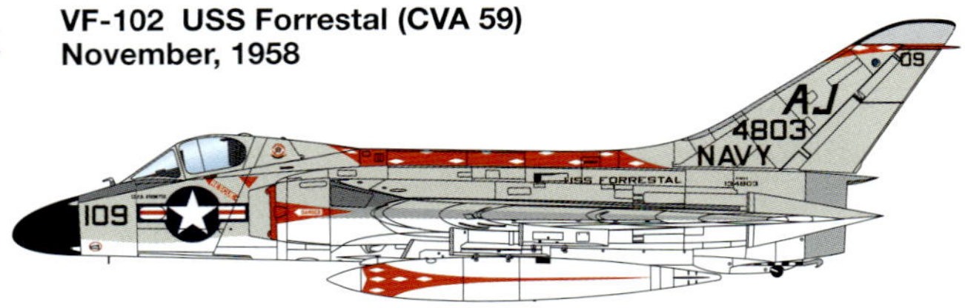

VF-102 USS Forrestal (CVA 59)
April, 1961

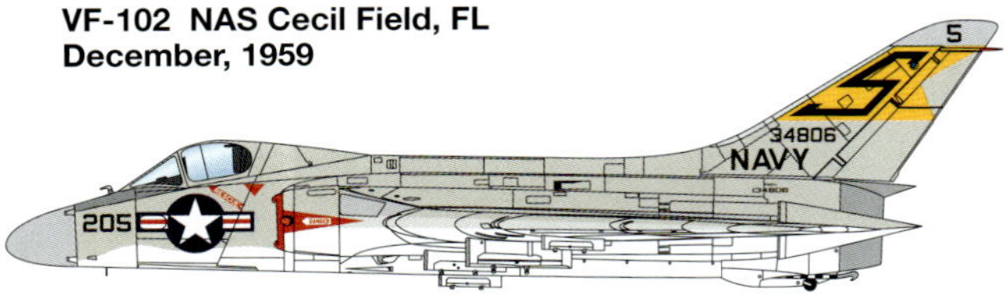

VF-141 NAS Miramar, CA
Summer, 1956

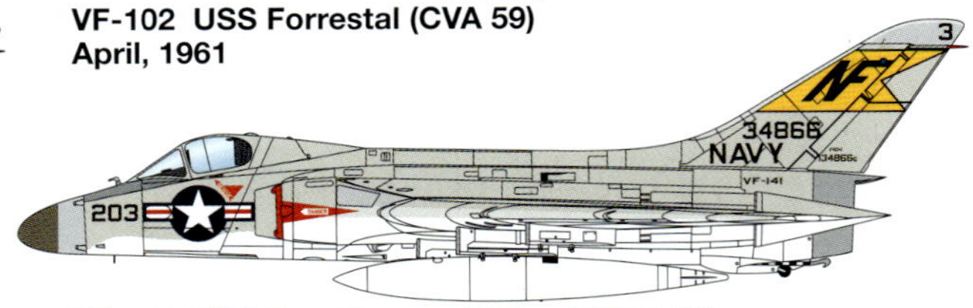

VF-141 USS Bon Homme Richard (CVA 31)
1957

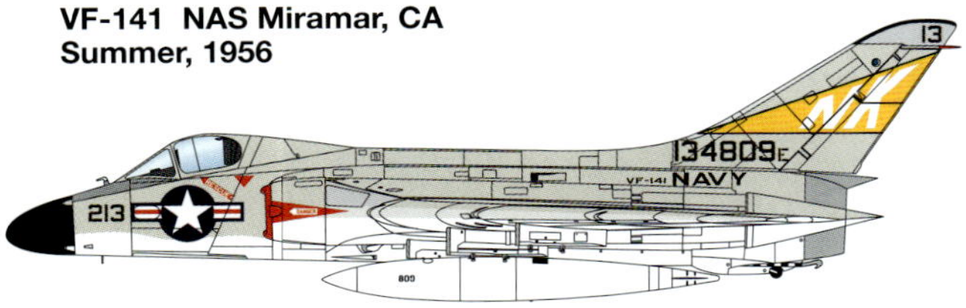

VF-141 NAS Miramar, CA
Summer, 1958

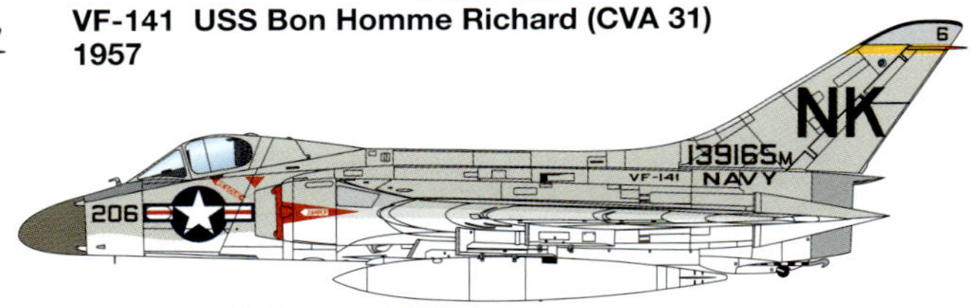

VF-141 USS Ranger (CVA 61)
Early 1959

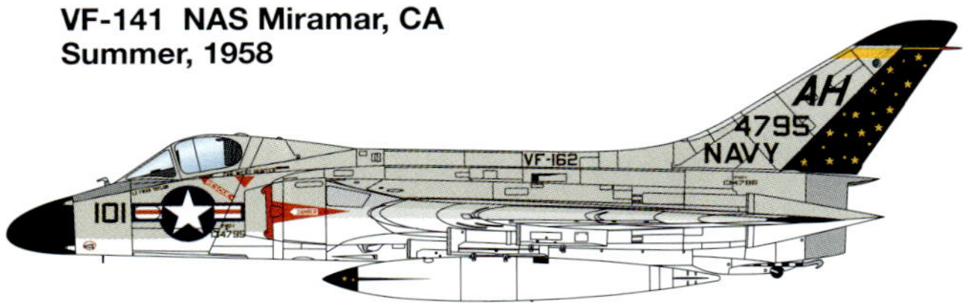

VF-162 NAS Cecil Field, FL
February, 1961

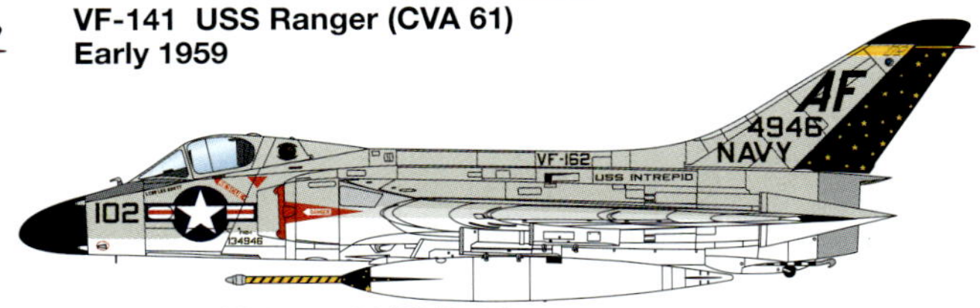

VF-162 USS Intrepid (CVA 11)
Late, 1961

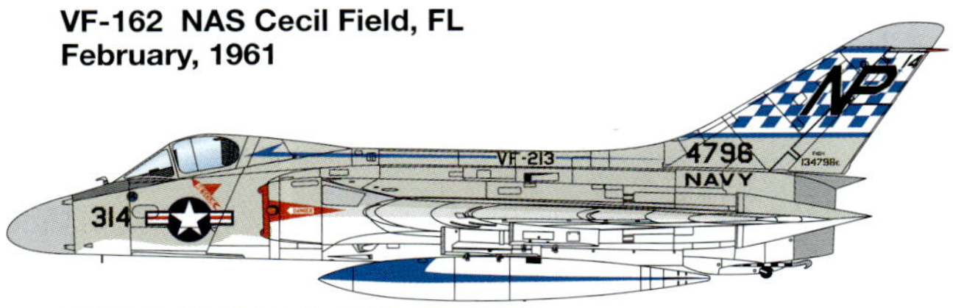

VF-213 NAS Moffett Field, CA
August, 1957

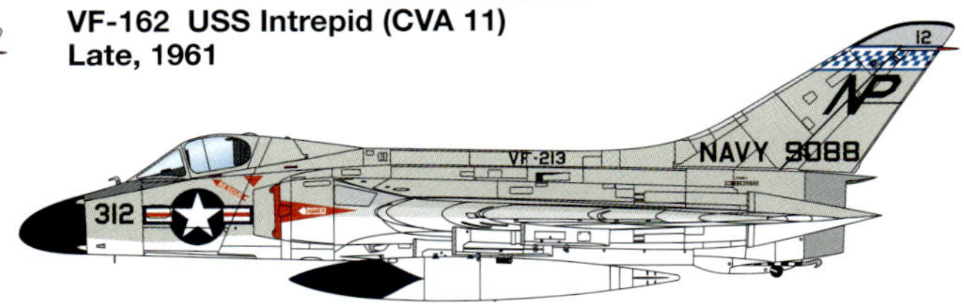

VF-213 NAS Moffett Field, CA
November, 1959

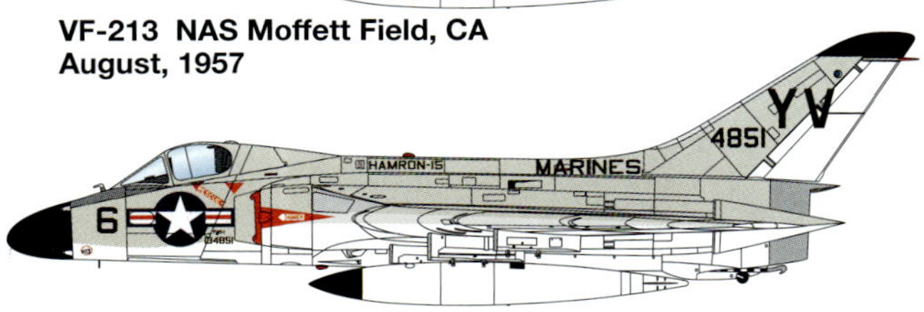

H&MS-15 MCAS El Toro, CA
1961

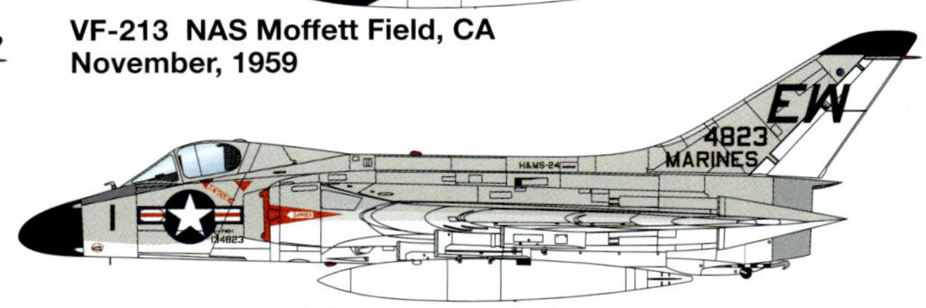

H&MS-24 MCAS Cherry Point, NC
1961

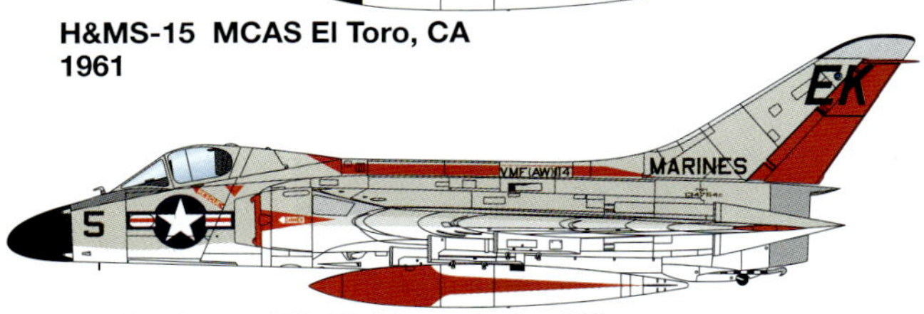

VMF(AW)-114 MCAS Cherry Point, NC
1958

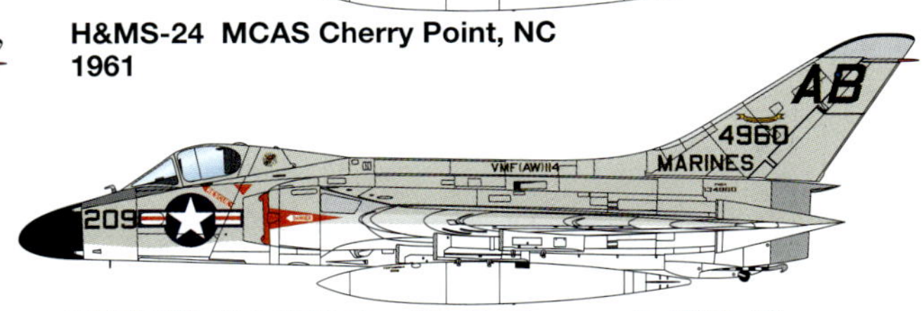

VMF(AW)-114 USS Franklin D. Roosevelt (CVA 42)
February, 1959

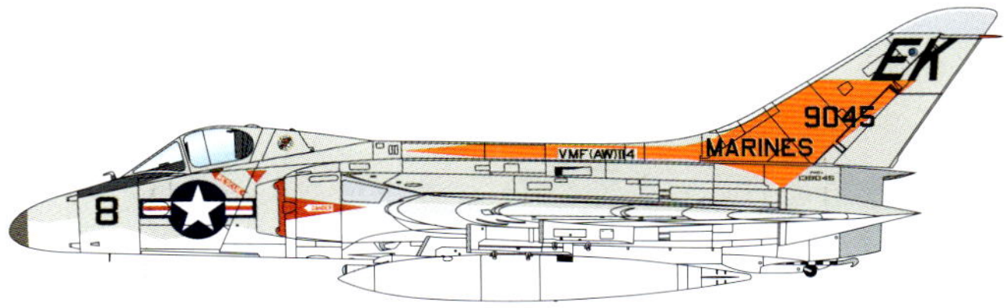

VMF(AW)-114 MCAS Cherry Point, NC
May, 1960

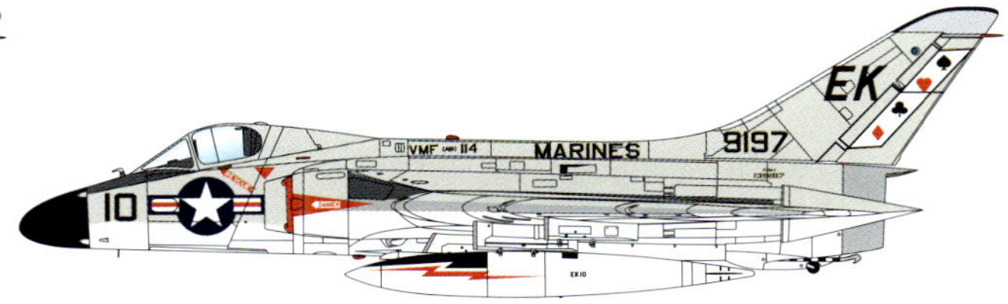

VMF(AW)-114 NAS Atsugi, Japan
August, 1962

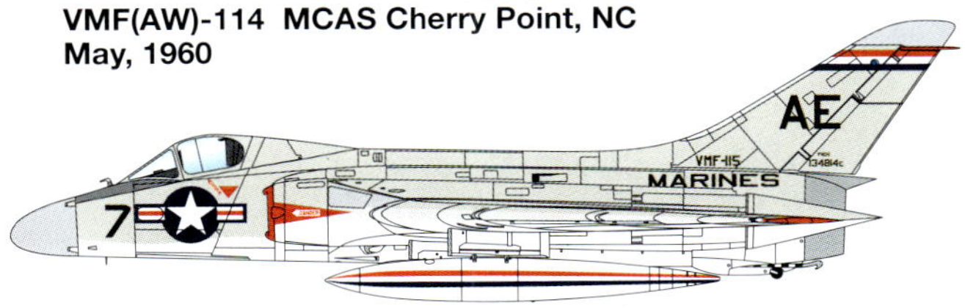

VMF(AW)-115 MCAS Cherry Point, NC
April, 1957

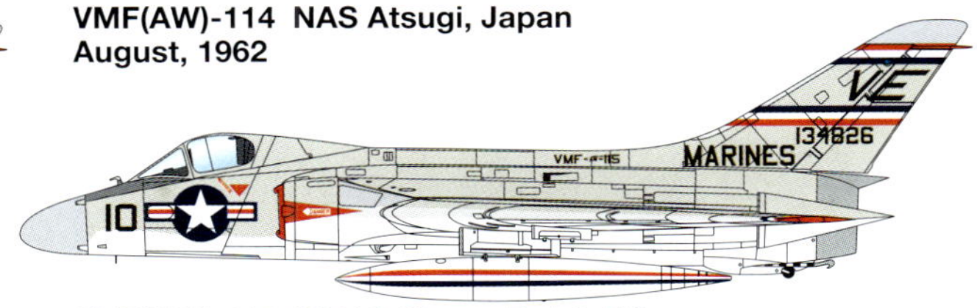

VMF(AW)-115 MCAS Cherry Point, NC
Summer, 1957

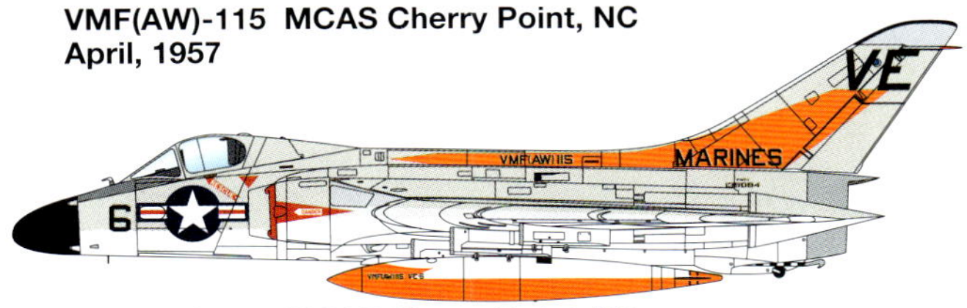

VMF(AW)-115 MCAS Cherry Point, NC
Spring, 1959

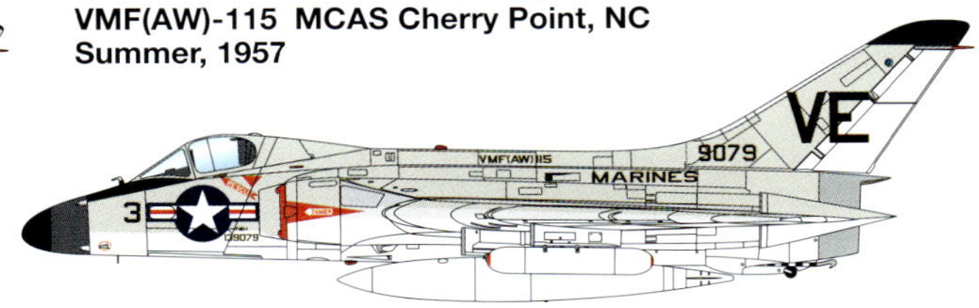

VMF(AW)-115 MCAS Cherry Point, NC
August, 1962

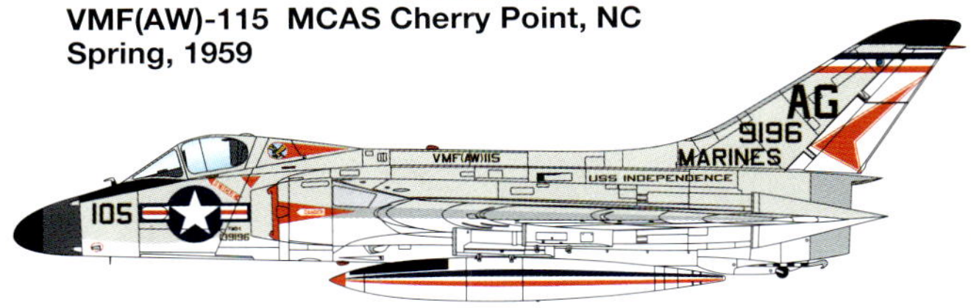

VMF(AW)-115 USS Independence (CVA 62)
March, 1963

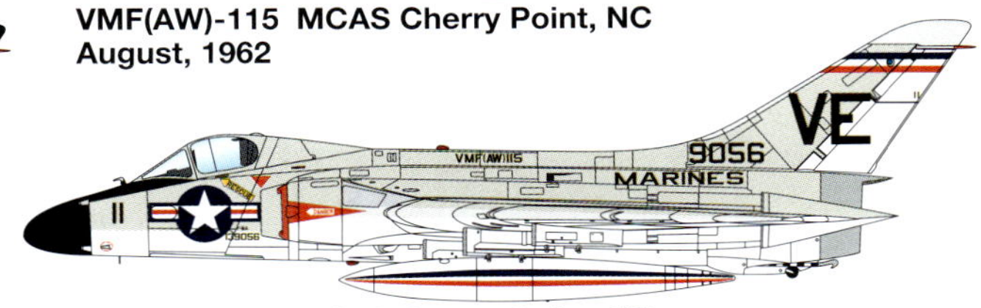

VMF(AW)-115 MCAS Cherry Point, NC
March, 1964

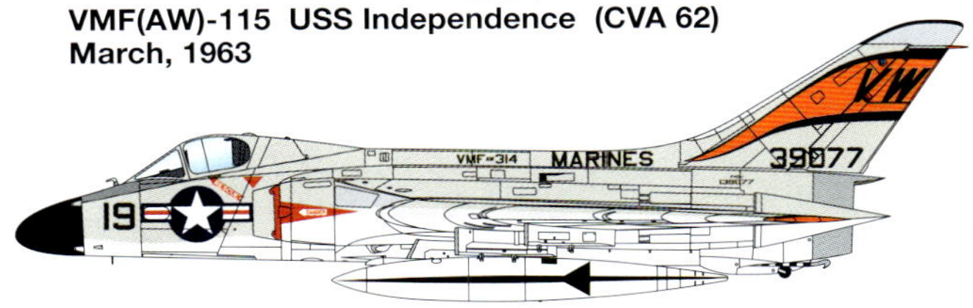

VMF(AW)-314 MCAS El Toro, CA
May, 1958

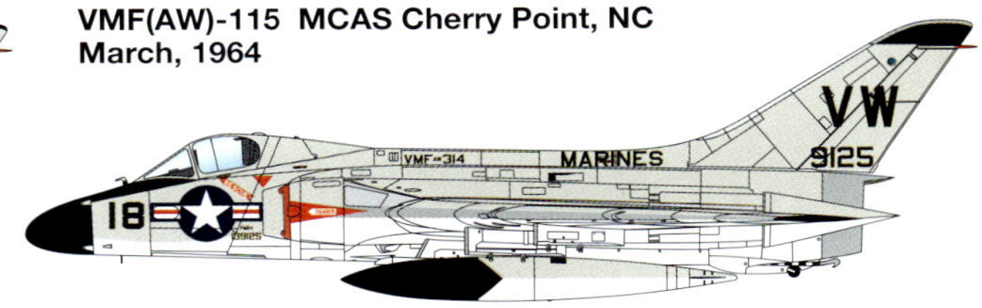

VMF(AW)-314 MCAS El Toro, CA
March, 1960

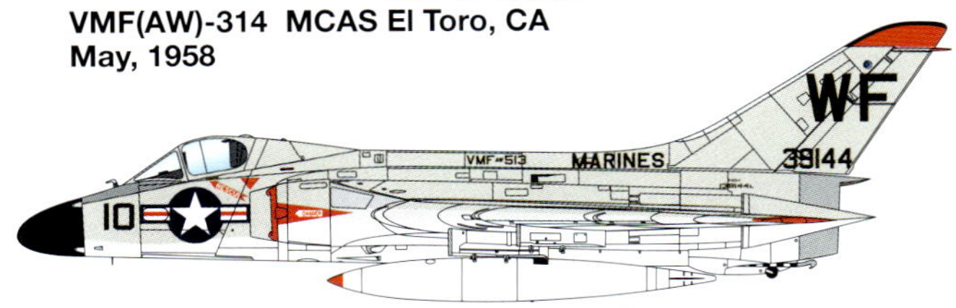

VMF(AW)-513 MCAS Cherry Point, NC
December, 1959

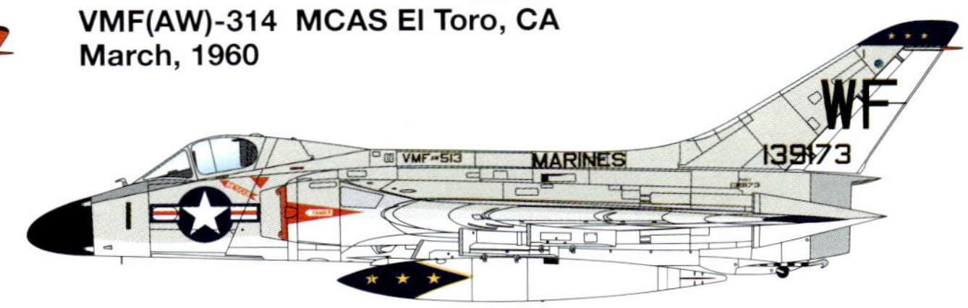

VMF(AW)-513 NAS Atsugi, Japan
December, 1961

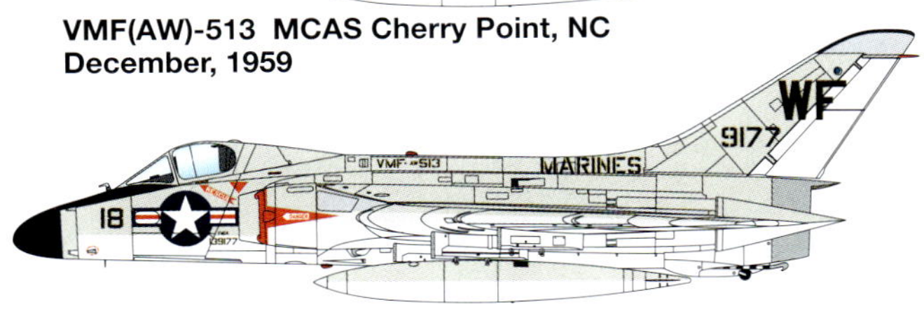

VMF(AW)-513 MCAS Cherry Point, NC
October, 1962

VMF(AW)-531 MCAS Cherry Point, NC
December, 1958

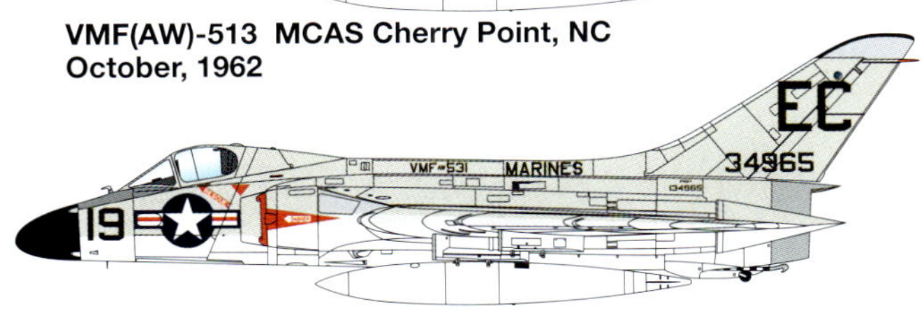

VMF(AW)-531 MCAS Cherry Point, NC
November, 1959

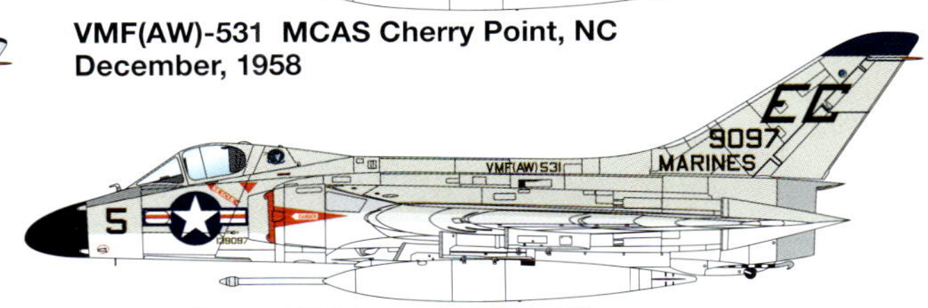

VMF(AW)-531 USS Intrepid (CVA 11)
Summer, 1960

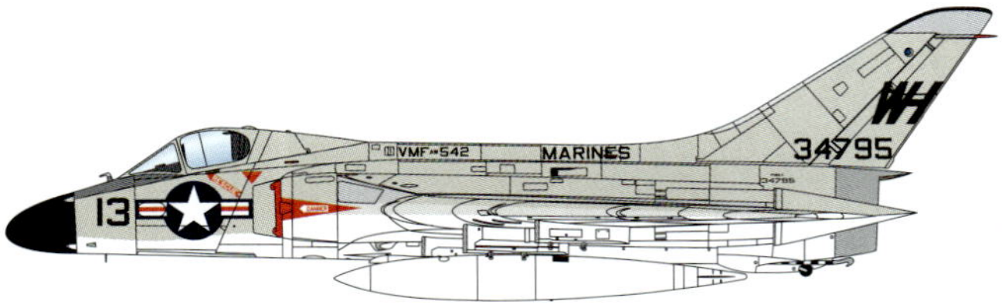

VMF(AW)-542 MCAS El Toro, CA
Early 1959

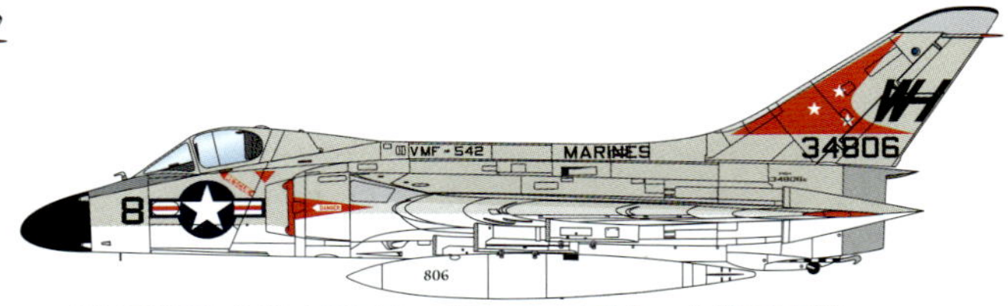

VMF(AW)-542 USS Bon Homme Richard (CVA 31)
July, 1959

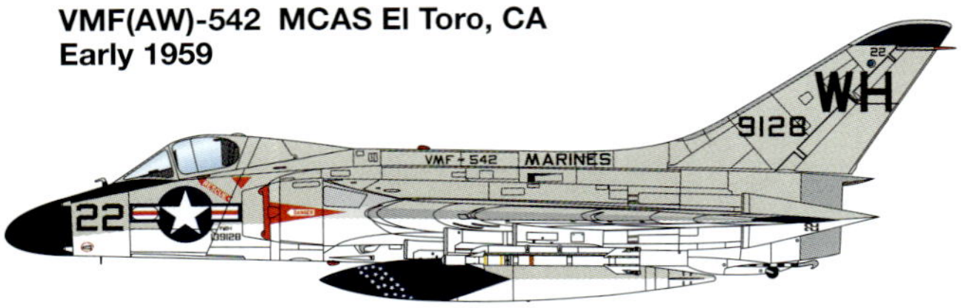

VMF(AW)-542 MCAS El Toro, CA
September, 1962

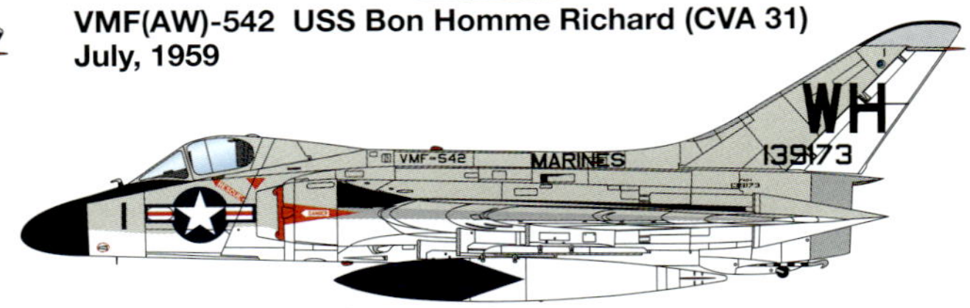

VMF(AW)-542 NAS Atsugi, Japan
December, 1962

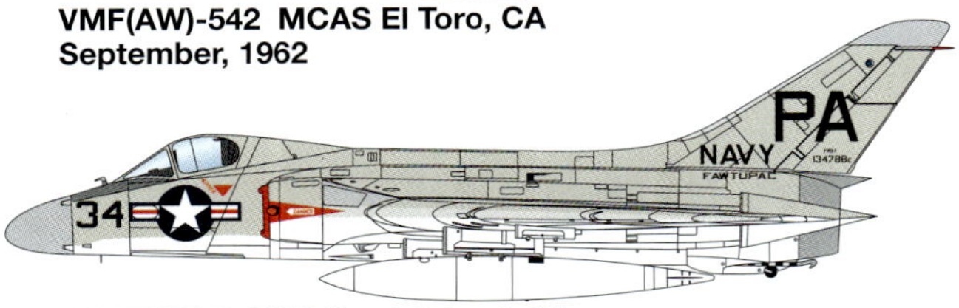

FAWTUPAC NAS North Island, CA
August, 1956

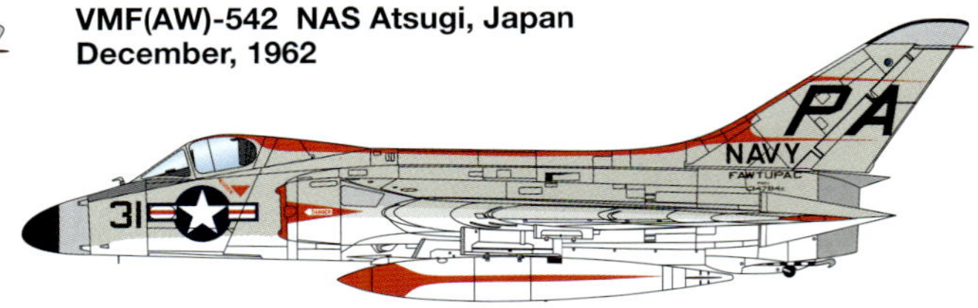

FAWTUPAC NAS North Island, CA
April, 1958

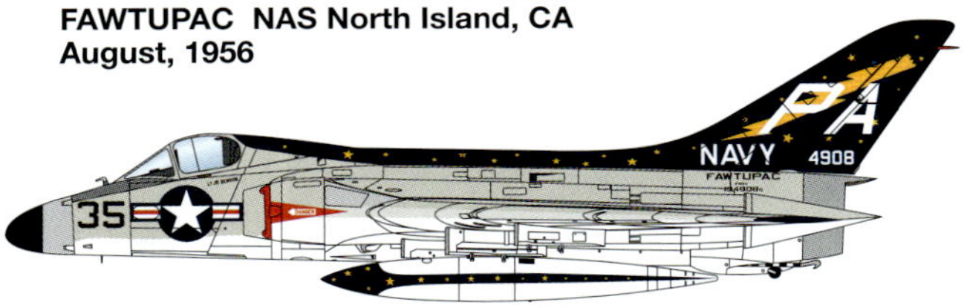

FAWTUPAC NAS North Island, CA
1958

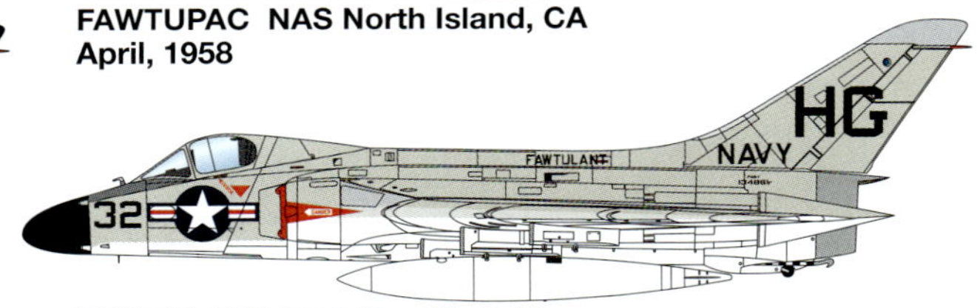

FAWTULANT, NAS Key West, FL
Early 1958

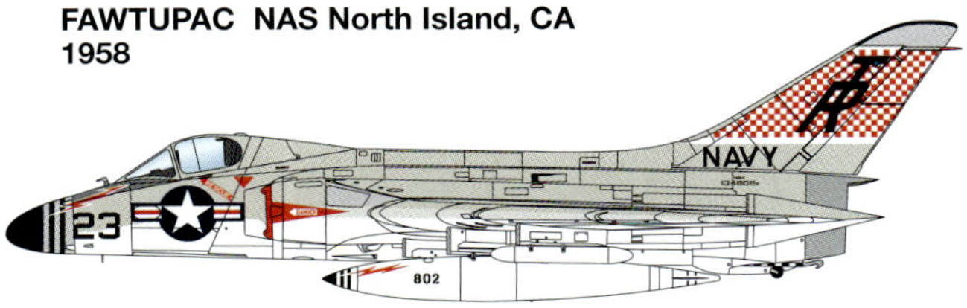

FAGU NAS El Centro, CA
February, 1958

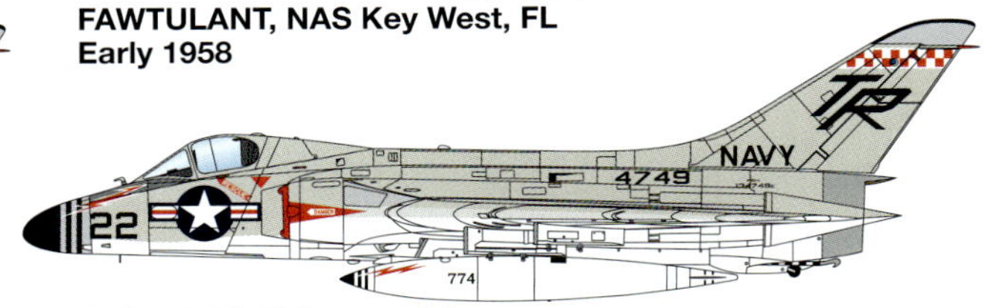

FAGU NAS El Centro, CA
Early 1959

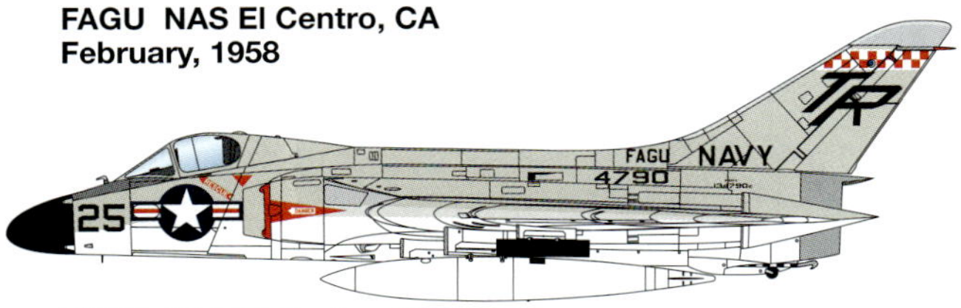

FAGU NAS El Centro, CA
December, 1959

VU-3 NAS North Island, CA
Spring, 1963

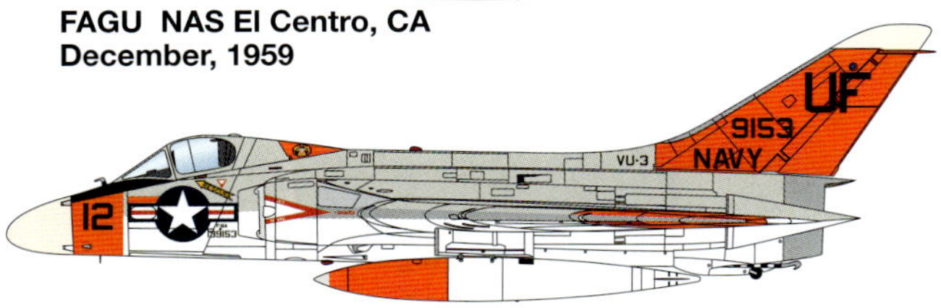

VU-3 NAS North Island, CA
May, 1964

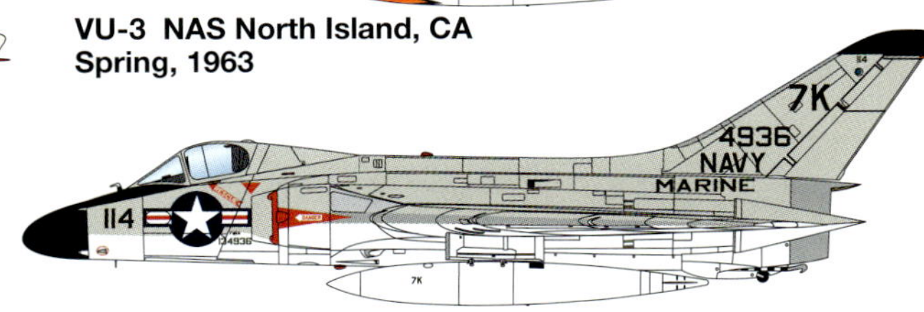

NARTU/MARTD NAS Olathe, KS
March, 1963

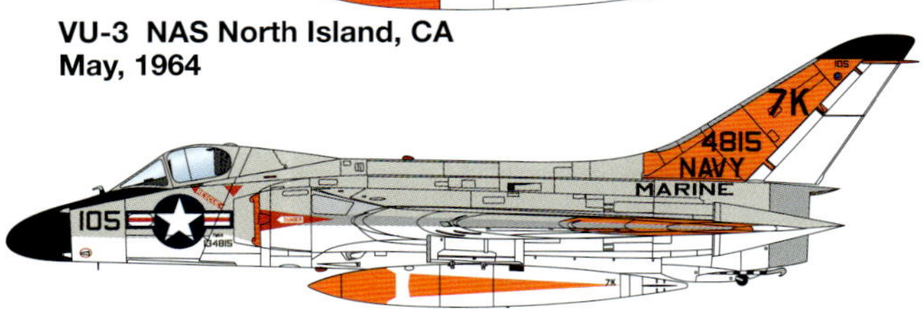

NARTU/MARTD NAS Olathe, KS
Summer, 1963

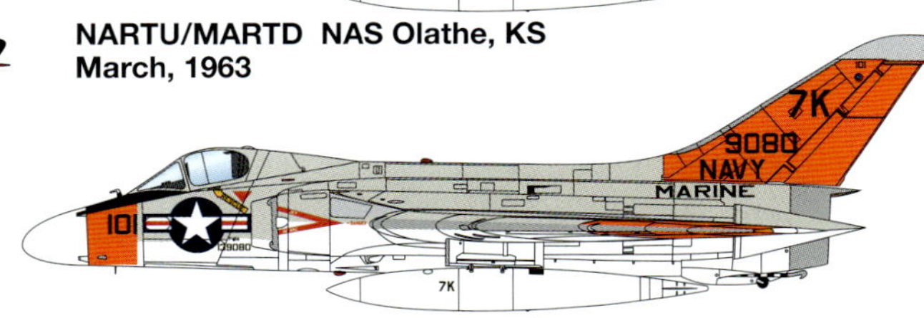

NARTU/MARTD NAS Olathe, KS
July, 1964

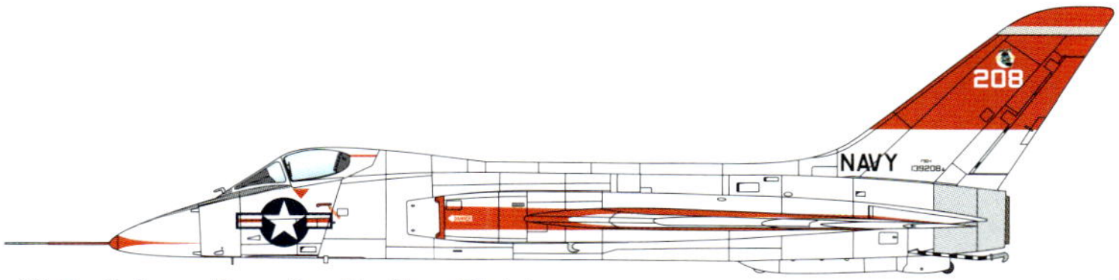

#1 Prototype, Douglas Testing Division
Edwards AFB, CA April, 1956

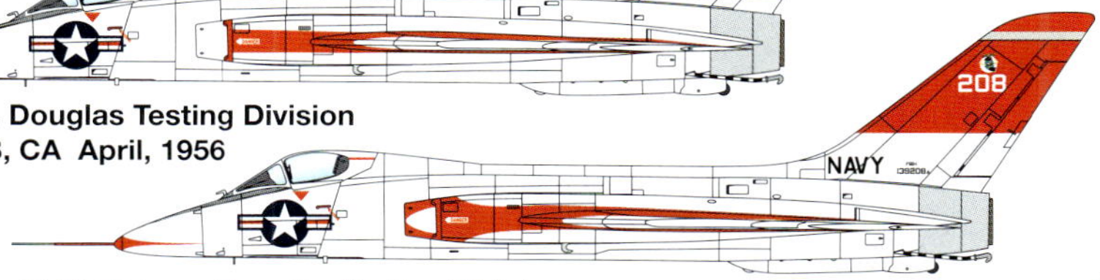

#1 Prototype, Douglas Testing Division
Edwards AFB, CA April, 1957

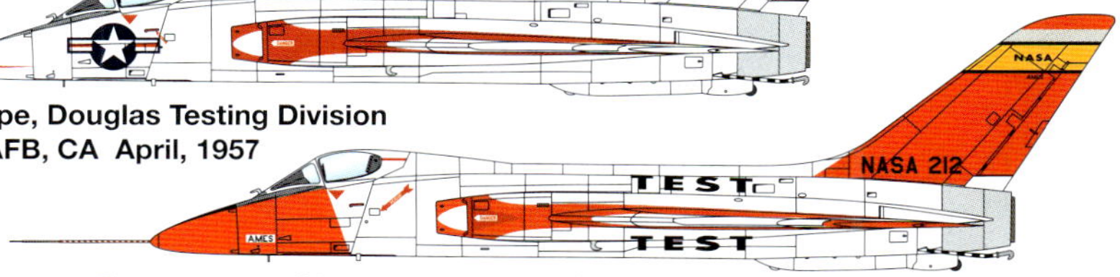

#1 Prototype, NASA Ames Research Center
NAS Moffett Field, CA April, 1960

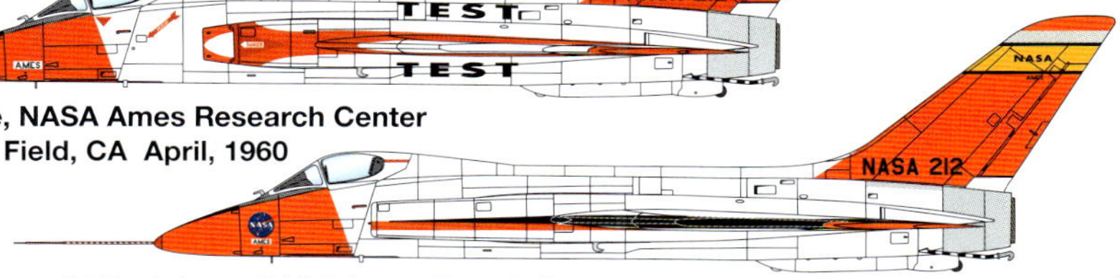

#1 Prototype, NASA Ames Rearch Center
NAS Moffett Field, CA 1965

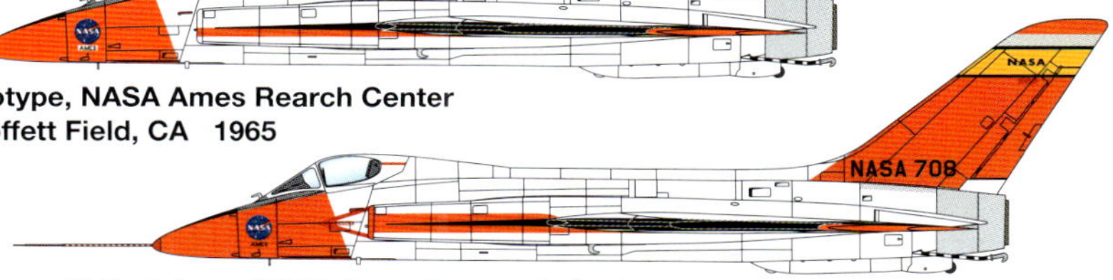

#1 Prototype, NASA Ames Research Center
NAS Moffett Field, CA October, 1967

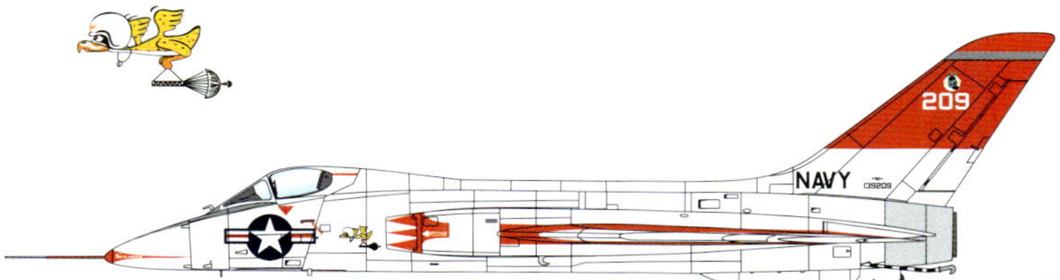

#2 Prototype, Douglas Testing Division
Edwards AFB, CA Summer, 1957

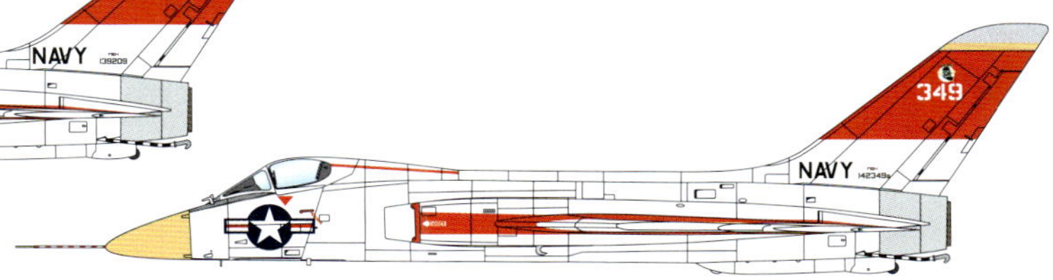

#3 Prototype, Douglas Testing Division
Edwards AFB, CA April, 1957

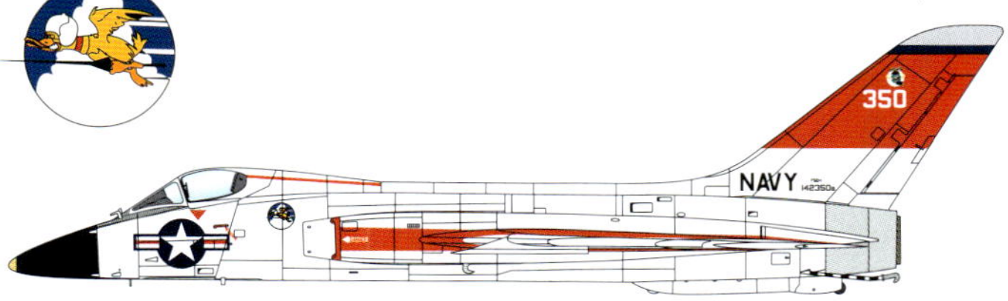

#4 Prototype, Douglas Testing Division
Edwards AFB, CA Summer, 1957

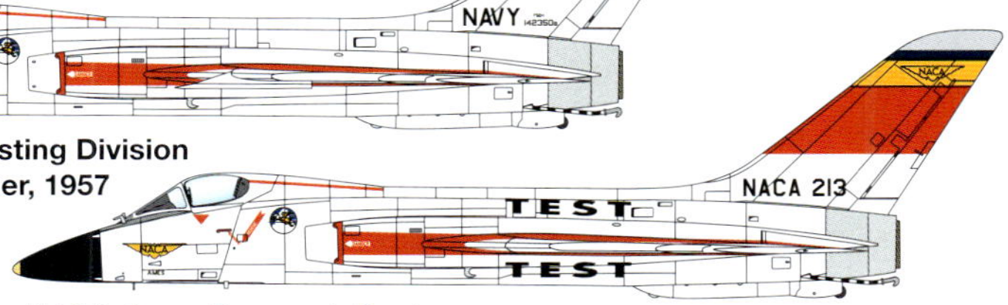

#4 Prototype, NACA Ames Research Center
NAS Moffett Field, CA September, 1967

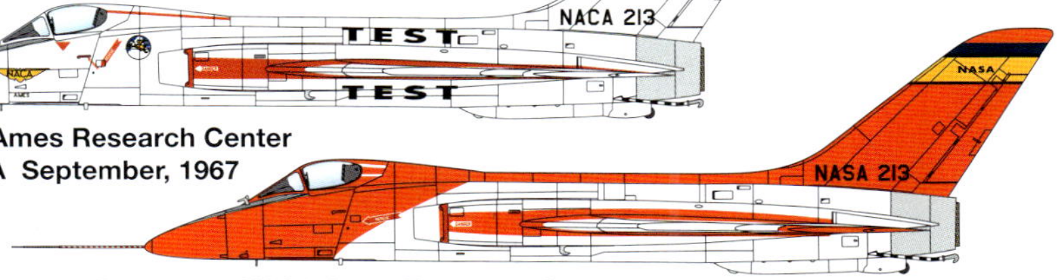

#4 Prototype, NASA Ames Research Center
Edwards AFB, CA 1962

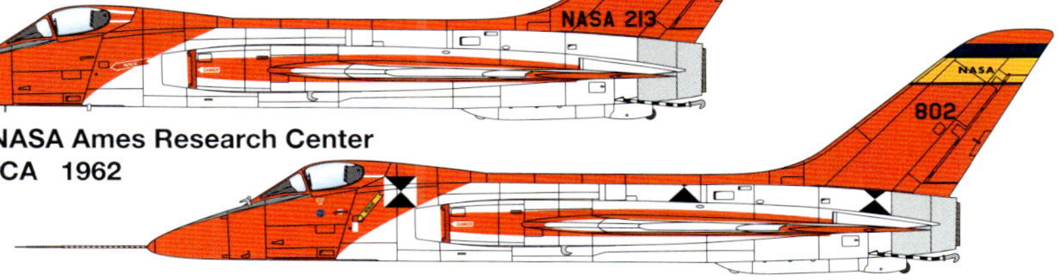

#4 Prototype, NASA Dryden Flight Research Center
Edwards AFB, CA 1967

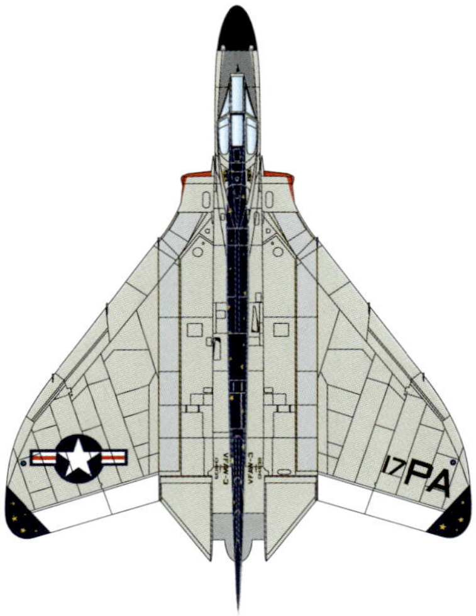

VF(AW)-3
NAS North Island, CA 1960

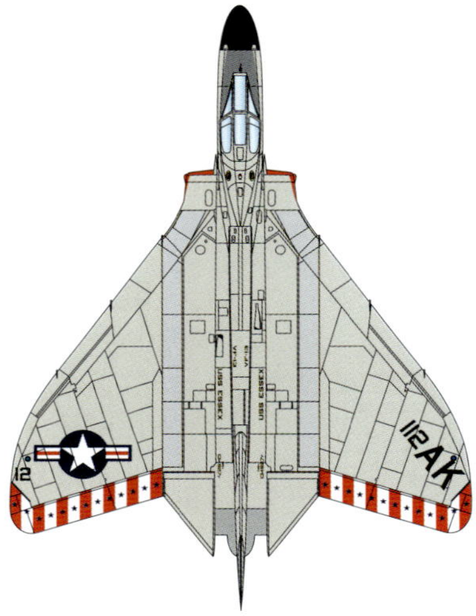

VF-13 USS Essex (CVA 9)
August, 1960

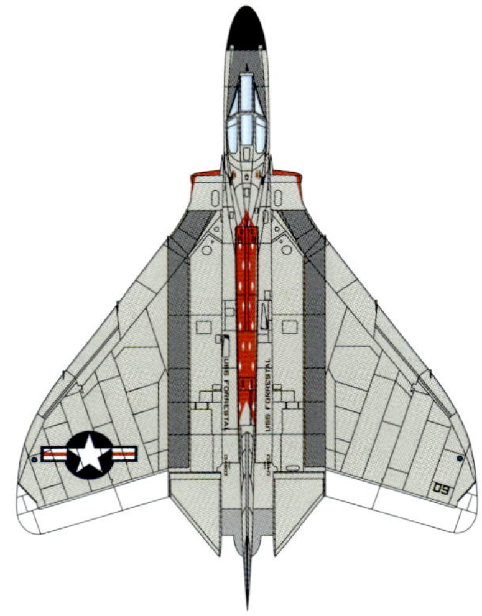

VF-102 USS Forrestal (CVA 59)
April, 1961

VF-162 USS Intrepid (CVA 11)
Late, 1961

VMF(AW)-114
MCAS Cherry Point, NC 1958

VMF(AW)-115
MCAS Cherry Point, NC April, 1957

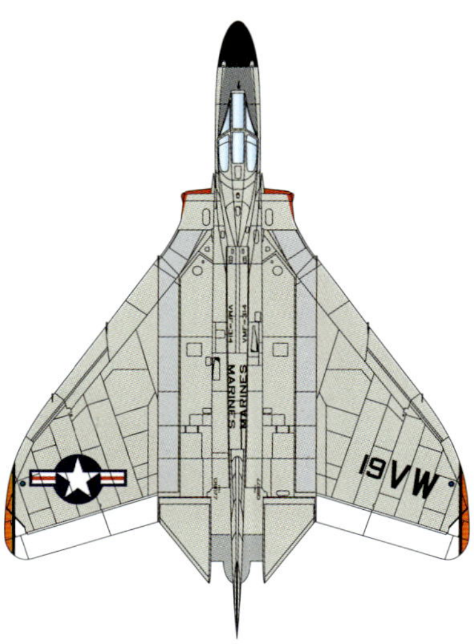

VMF(AW)-314
MCAS El Toro, CA May, 1958

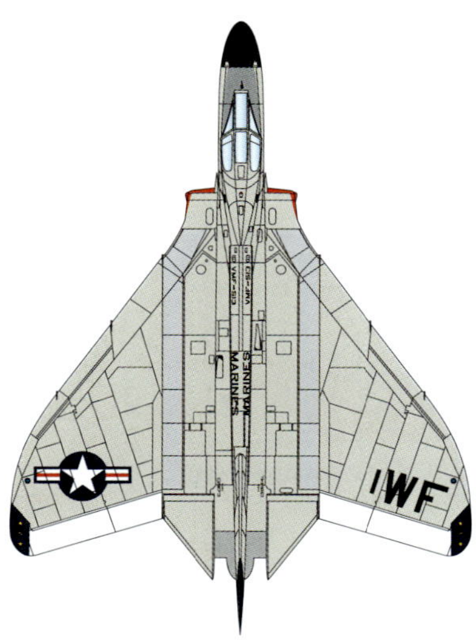

VMF(AW)-513
NAS Atsugi, Japan December, 1961

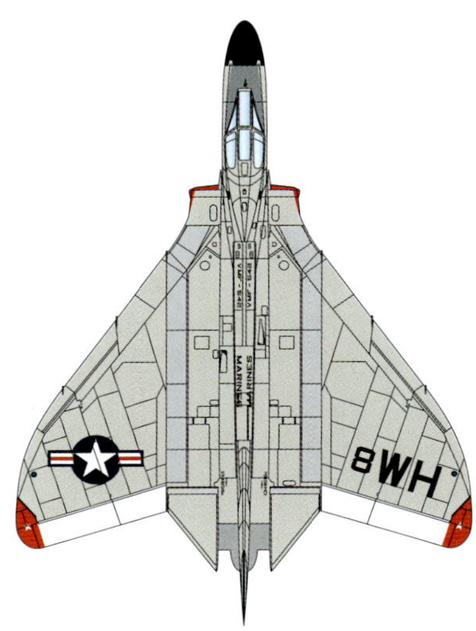

VMF(AW)-542 USS Bon Homme
Richard (CVA 31) July, 1959

Fleet All Weather Training Unit Pacific
NAS North Island, CA April, 1958

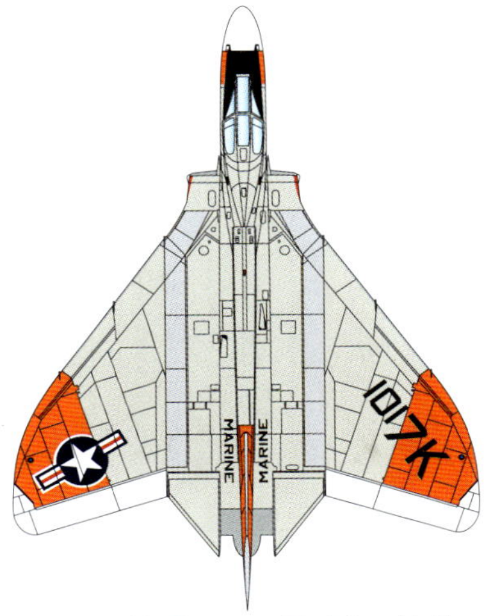

Naval Air Reserve Training Unit
NAS Olathe, KS July, 1964

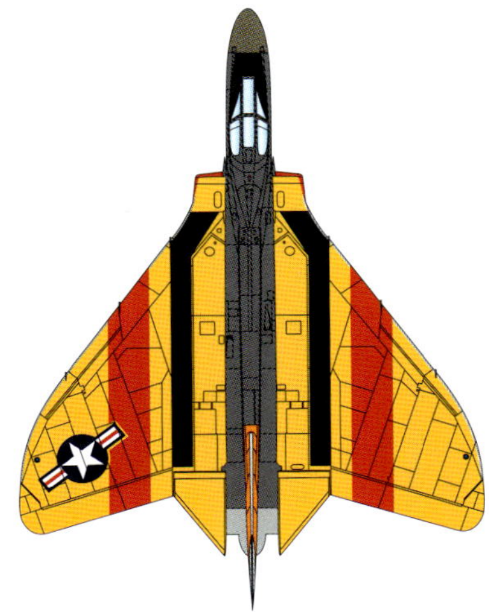

Naval Air Test Center
NAS Patuxent River, MD May, 1960

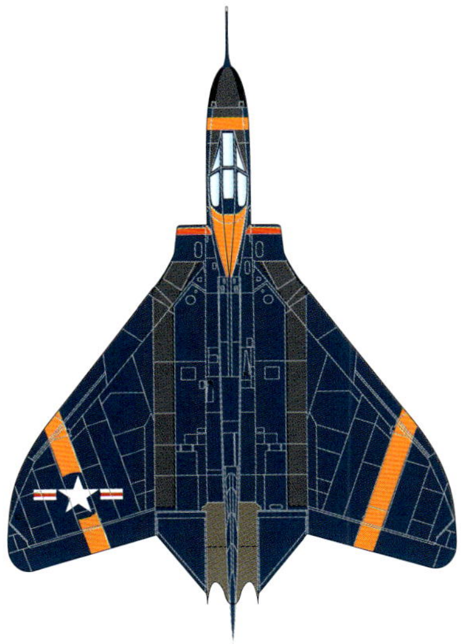

#1 Prototype Douglas Testing Division
NAF China Lake, CA 1955

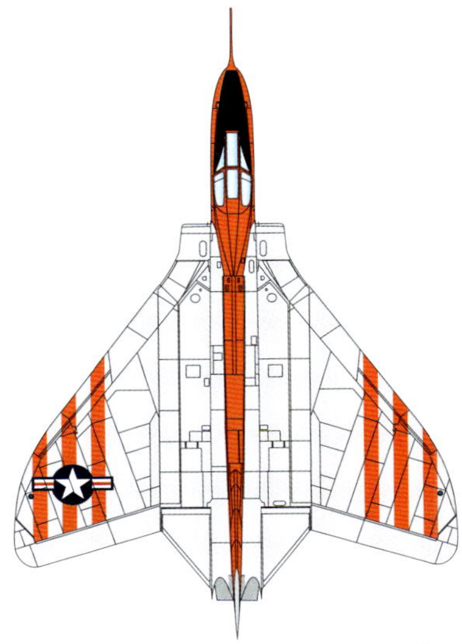

#2 Prototype/GE CJ-805 Engine Test
Edwards AFB, CA April, 1958

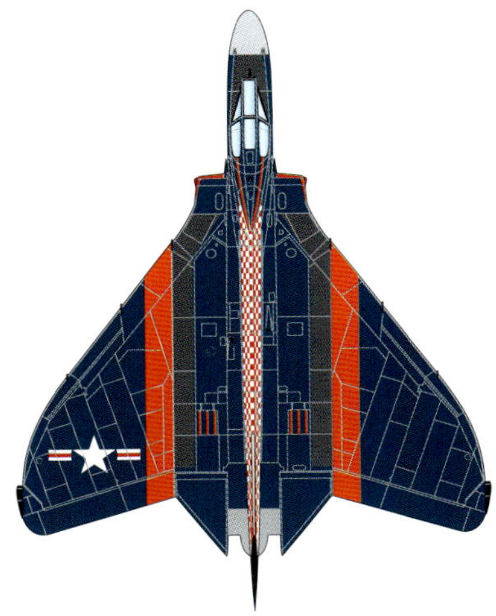

Douglas Testing Division
NAF China Lake, CA May, 1957

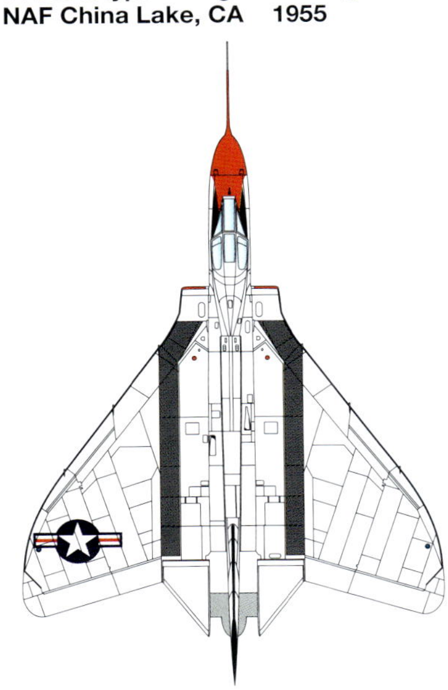

Douglas Testing Division
Edwards AFB, CA Summer, 1954

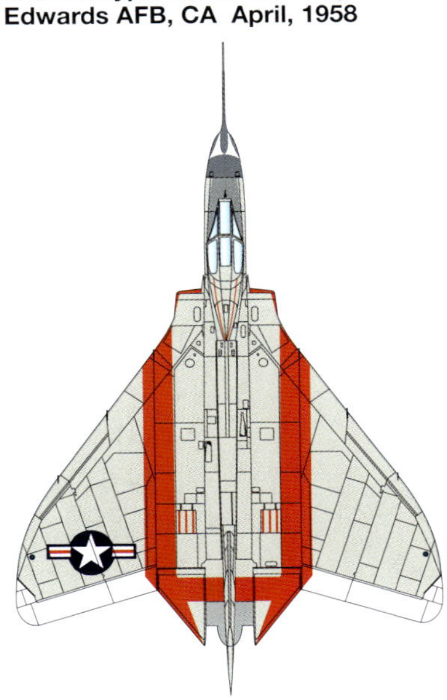

Douglas Testing Division
NAS Point Mugu, CA May, 1958

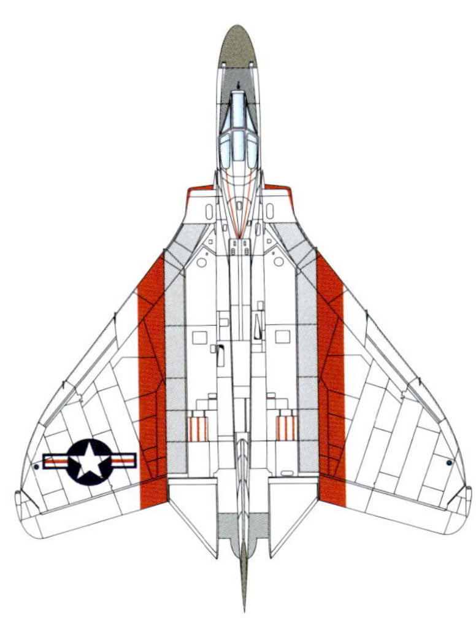

Douglas Testing Division
NAF China Lake, CA August, 1956

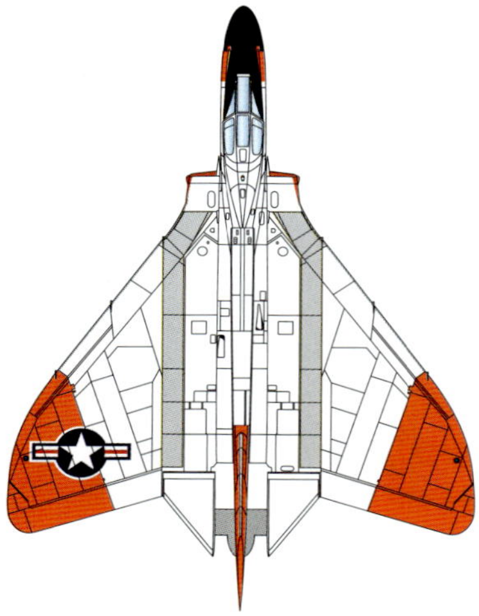

Naval Air Development Unit
NAS South Weymouth, MA April, 1960

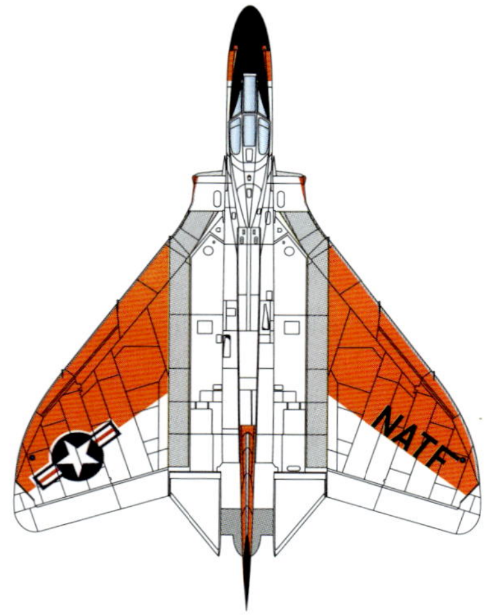

Naval Air Test Facility
NAS Lakehurst, NJ May, 1961

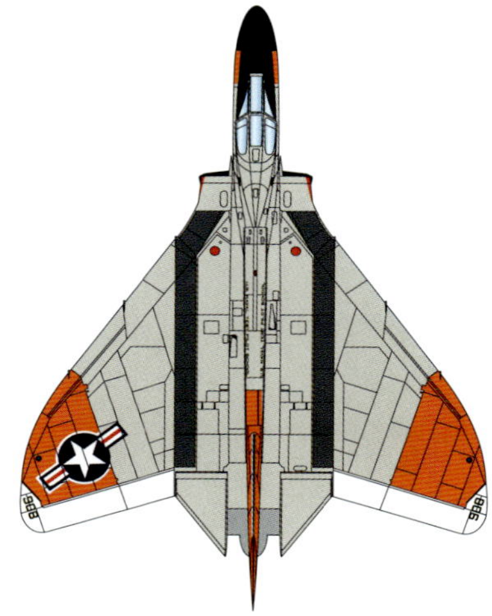

US Naval Test Pilot School
NAS Patuxent River, MD 1967

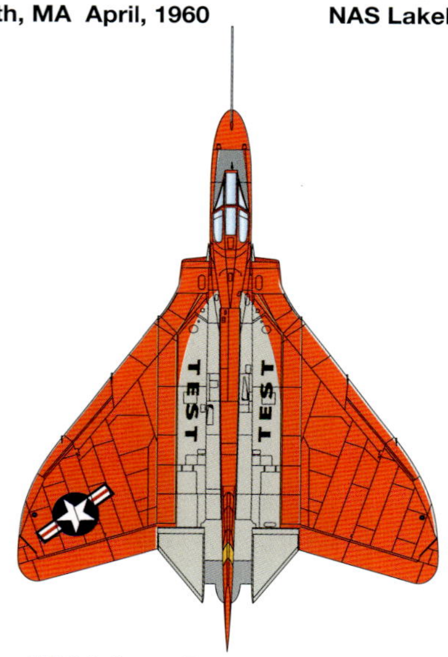

NASA Ames Research Center
NAS Moffett Field, CA May, 1959

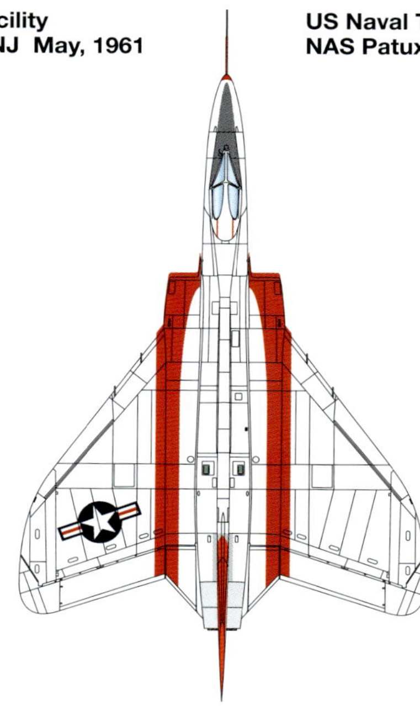

Douglas Testing Division
Edwards AFB, CA April, 1957

NASA Ames Research Center
NAS Moffett Field, CA 1965

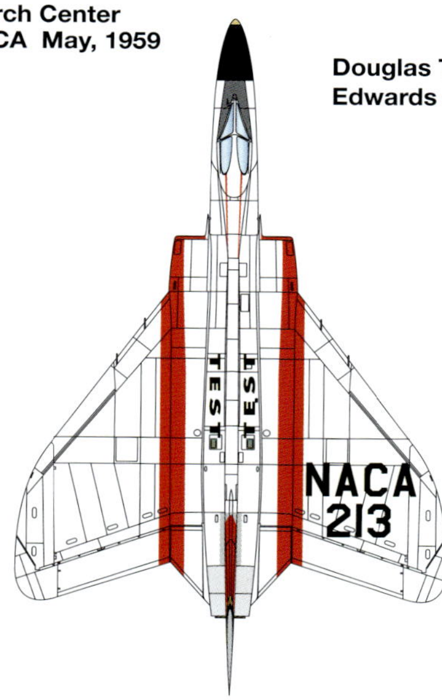

NASA Ames Research Center
NAS Moffett Field, CA 1957

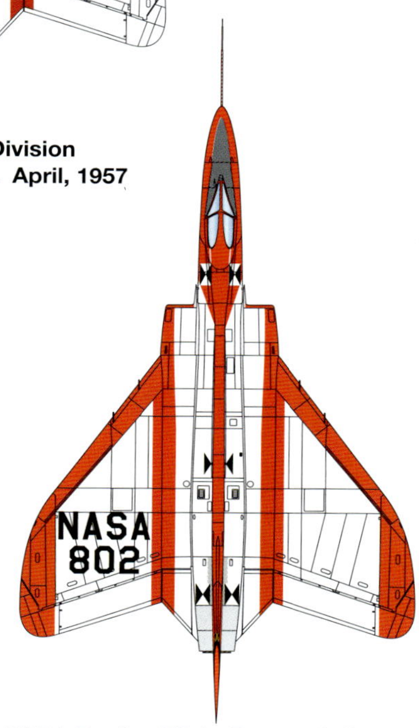

NASA Dryden Flight Research Center
Edwards AFB, CA 1967

BIBLIOGRAPHY

Babcock, Elizabeth, *History of the Navy at China Lake, Volume 3, Magnificent Mavericks* (Washington, D.C.: Naval Historical Center, 2008), pp. 413–416

Bradley, Robert E., AAHS Journal Vol. 48 No. 4 Winter 2003, "Flight Report No. 1, Model XF4D-1," Douglas Aircraft Co., Testing Division, April 1951, p.256

Bowman, Martin W. with Matthias Vogelsang, Lockheed F-104 Starfighter, (Ramsbury, Marlborough Wiltshire, United Kingdom: Crowood Press Ltd., 2000), p. 21

Cebeci, Tuncer, *Legacy of a Gentle Genius The Life Of A.M.O. Smith* (Long Beach, California: Horizons Publishing, 1999), p.30

Dalquest, Falk, Haskell, McGinnis, Schoneberger and Travers, *Seven Decades of Progress A Heritage of Aircraft Turbine Technology* (Fallbrook, CA: Aero Publishers, Inc., 1979), p. 93

Gill, W.F. with John Brizendine, Report "Contractor Demonstration of the Carrier Suitability of the XF4D-1" by W.F. Gill and J.C. Brizendine. Douglas Aircraft Co. Testing Division, May 24, 1954

Ginter, Steve, *Naval Fighters Number Thirty-five,* Douglas F5D-1 Skylancer (Simi Valley, CA: Steve Ginter, 1996), p.4

Godwin, Robert, *Dyna-Soar – Hypersonic Strategic Weapons System* (Burlington, Ontario, Canada: Collector's Guide Publishing Inc., 2003) pp.224–226

Gunston, Bill, *The Development of Jet and Turbine Aero Engines* (Sparkford, Nr Yeovil, Somerset, U.K.: Patrick Stephens, Ltd., 1995), p.153

Gunston, Bill, *Early Supersonic Fighters of the West* (London, Ian Allan Ltd., 1976) pp.230–231

Gunston, William, *Early Supersonic Fighters of the West* (Shepperton, Surry, United Kingdom: Ian Allan Ltd., 1976) pp.215, 217

Heinemann, Edward H., and Rosario Rausa, *Ed Heinemann Combat Aircraft Designer* (Annapolis MD: United States Naval Institute, 1980) pp. 187, 195–196

Heinemann, Edward H., "A Veteran Designer's Thoughts on Fighters" *Air International,* October 1982, pp.180, 181

Jenkins, Dennis R. and Tony R. Landis, *Experimental and Prototype U.S. Air Force Jet Fighters* (North Branch, MN: Specialty Press, 2008) pp.122–124

Kay, Antony L., *Turbojet History and Development 1930–1960* (Ramsbury, Marlborough Wiltshire SN8 2HR, United Kingdom:The Crowood Press Ltd., 2007), p.114

KLD Associates, "A Review of the Flight Characteristics of the F4D-1 and F7U-3 Aircraft with Particular Emphasis on Carrier Suitability" Final Report, July 7, 1986

Lagasse, Paul D., *Master of Arts Thesis, The Westinghouse Aviation Gas Turbine Division 1950–1960: A Case Study Of The Role Of Failure In Technology And Business* (1997), p.19–23

Markman, Steve, and Bill Holder, *One-Of-A-Kind Research Aircraft, a History of In-Flight Simulators, Testbeds & Prototypes* (Atgen, PA: Schiffer Publishing Ltd., 1995), p.66

Matranga, Gene J., William H. Dana, and Neil A. Armstrong, "Flight-Simulated Off-The-Pad escape and Landing Maneuvers For a Vertically Launched Hypersonic Glider" NASA Technical Memorandum X-637.

Meyer, Corwin Corky, *Naval Fighters Number Twenty Six, Grumman Swing-Wing XF10F-1 Jaguar* (Simi Valley, CA, Ginter, 1993), p.20–21

Mutza, Wayne, *Convair F-102 Delta Dagger* (Atglen, PA: Schiffer Publishing Ltd. 1999) pp.56–71

Oestrreicher, Philip F., Flight Report, "An Investigation of the Maneuvering Stability of the F-6A Airplane," NATC Patuxent River, MD, dated 10 August, 1966, and NATC Report of Test Results, "Investigation of the Transonic Flying Qualities of the F-6A Airplane," by P.F. Oestreicher, NATC Patuxent River, MD, dated 12 September, 1966

O'Rourke, Jerry, "The Douglas Navy F4D Ford," *AAHS Journal,* Summer 1979, pp.125–131. This material also appeared in a slightly different form, "The F4D 'Ford': A Better Idea?" Naval Institute Proceedings, June 1986, pp.123–126

Pape, Garry R. with John M. Campbell, *Northrop Flying Wings, A history of Jack Northrop's Visionary Aircraft* (Atglen, Pennsylvania: Schiffer Publishing Ltd.,1995), p.18

Pearcy, Arthur, *Flying The Frontiers NACA and NASA Experimental Aircraft* (Annapolis, Maryland: Naval Institute Press, 1993) pp.60–61

Rahn, Bob, with Rosario Rausa, *Tempting Fate* (North Branch MN: Specialty Press, 1997), p.149

Report "Preliminary Evaluation of Model XF4D-1 Airplane" Naval Air Test Center, Patuxent River, MD., Aug. 15, 1952

Report "Model XF4D-1 Airplane Carrier Suitability Tests Letter Report No. 1, Final Report" Naval Air Test Center, Patuxent River, Maryland, Flight Test Division, Jan 25, 1954

Report "Pilots' Evaluation of the XF4D-1 Skyray" Headquarters, Air Force flight Test Center, Edwards Air Force Base, California, 3 December 1953

Rolls, L. Stewart, with David G. Koenig, and Fred J. Drinkwater III, "Flight Investigation of the Aerodynamic Properties of an Ogee Wing" NASA Technical Note, NASA TN D-3071, December 1965

Spangenberg, Walter, with Rosario Rausa, Oral History (http://www.georgespangenberg.com/history1.htm), Tape 7 of 16, side A

Sullivan, Mark P., *Dependable Engines: The Story of Pratt & Whitney* (Reston, VA: American Institute of Aeronautics and Astronautics, Inc., 2008) pp.2–7

Thomason, Tommy H., *U.S. Naval Air Superiority: Development of Shipborne Jet Fighters 1943–1962* (North Branch, MN: Specialty Press, 2007), p.17, 92

Utility Flight Handbook for Navy Model F5D-1 Aircraft, 1 July 1957.

Williams, Nicholas, "The X-Rays Douglas' XF4D-1 Skyrays" *AAHS Journal,* Winter, 1977, p.245, 249

Williams, Nicholas, "Silver Eagle Skyrays" *The Hook,* Spring 1978, pp.14–17.

Williams, Nicholas, "The Douglas Skyray in Fleet Colors, Part I" *AAHS Journal,* Summer 1978, pp.136–137

Williams, Nicholas, "The Douglas Skyray in Fleet Colors, Part II" *AAHS Journal,* Summer 1981, pp. 100 –101.

Williams, Nicholas, "The Douglas Skyray in Fleet Colors, Part III" *AAHS Journal,* Summer 1985 pp.82–85

Willaims, Nicholas, "The Mayday Skyrays Douglas Testing Division F4D-1s 1954–1959" *AAHS Journal,* Summer 1986, p. 84

Yenne, Bill, *Convair Deltas: From SeaDart to Hustler* (North Branch, MN: Specialty Press, 2009) pp.119–125

ENDNOTES

Chapter One

1. Garry R. Pape with John M. Campbell, Northrop Flying Wings, A History of Jack Northrop's Visionary Aircraft (Atglen, Pennsylvania: Schiffer Publishing Ltd., 1995) p.18
2. Tuncer Cebeci, Legacy of a Gentle Genius: The Life of A.M.O. Smith (Long Beach, California: Horizons Publishing, 1999) p.30
3. Ibid. pp.30 & 35
4. George A. Spangenberg with Rosario Rausa, Oral History (http://www.georgespangenberg.com/history1.htm), Tape 7 of 16, side A
5. Edward H. Heinemann and Rosario Rausa, Ed Heinemann Combat Aircraft Designer (Annapolis, Maryland: Naval Institute Press, 1980) p.187
6. Nicholas M. Williams, The X-rays Douglas' XF4D-1 Skyrays, Journal, Winter 1977 p.245
7. Heinemann, Combat Aircraft Designer, p.187
8. Ibid. p.188
9. William Gunston, Early Supersonic Fighters of the West (Shepperton, Surry, United Kingdom: Ian Allan Ltd., 1976) p.217
10. Mock-up Inspection Report, Model XF4D-1 Airplane
11. Spangenberg Oral History, Tape 4 of 16, side A Side Bar 1
12. Dennis R. Jenkins and Tony R. Landis, Experimental and Prototype U.S. Air Force Jet Fighters (North Branch, Minnesota: Specialty Press, 2008) pp.122-124
13. Robert E. Bradley AAHS Journal Vol. 48, No. 4 Winter 2003, p.256

Chapter Two

1. "Flight Report No. 1, Model XF4D-1," Douglas Aircraft Co., Testing Division, Apr. 1951
2. Heinemann and Rausa, Ed Heinemann Combat Aircraft Designer p.189
3. Bob Rahn with Rosario Rausa, Tempting Fate (North Branch, Minnesota: Specialty Press, 1997) p.83
4. Ibid. p.85
5. Ibid. p.85
6. Ibid. pp.86-87
7. Report "Preliminary Evaluation of Model XF4D-1 Airplane" Naval Air Test Center, Patuxent River, Maryland, Aug. 15, 1952
8. Ibid.
9. Report "Contractor Demonstration of the Carrier Suitability of the XF4D-1" by W. F. Gill and J. C. Brizendine. Douglas Aircraft Co., Testing Division, May 24, 1954.
10. Taped Interview of Marshall V. Beebe, CDR USN (Ret.) conducted by Nick Williams, Aug 31, 1976.
11. Ibid.
12. Report "Model XF4D-1 Airplane Carrier Suitability Tests Letter Report No. 1, Final Report" Naval Air Test Center, Patuxent River, Maryland, Flight Test Division, Jan. 25, 1954.
13. Rahn "Tempting Fate" p.94
14. "YF-100 Records" Aviation Week, November 2, 1953 p.16; "F-100, F4D Battle for Top Speed Laurels," Aviation Week, November 9, 1953. p.14
15. Rahn "Tempting Fate" p.94
16. Ibid. p.103
17. Report "Pilots' Evaluation of the XF4D-1 Skyray" Headquarters, Air Force Flight Test Center, Edwards Air Force Base, California, 3 December 1953
18. Dalquest, Falk, Haskell, McGinnis, Schoneberger and Travers, Seven Decades of Progress: A Heritage of Aircraft Turbine Technology, (Fallbrook, California: Aero Publishers, Inc., 1979) p.93
19. Taped Interview of Bill Todd, General Electric Test Pilot, conducted by Nick Williams, Sept. 8, 1976
20. Ibid.
21. Jenkins and Landis, Experimental & Prototype U.S. Air Force Jet Fighters. pp.125-128

Chapter Three

1. Paul D. Lagasse, Master of Arts Thesis, The Westinghouse Aviation Gas Turbine Division 1950-1960: A Case Study Of The Role Of Failure In Technology And Business (1997) pp.19-23
2. Tommy Thomason, U.S. Naval Air Superiority: Development of Shipborne Jet Fighters 1943-1962 (North Branch, Minnesota: Specialty Press, 2007) p.92
3. Ibid. p.93
4. Corwin Corky Meyer, Naval Fighters Number Twenty-Six, Grumman Swing-Wing XF10F-1 Jaguar (Simi Valley, California, Steve Ginter, 1993) pp.20-21
5. Antony L. Kay, Turbojet History and Development 1930-1960 (Ramsbury, Marlborough Wiltshire SN8 2HR, United Kingdom: The Crowood Press Ltd., 2007) p.114
6. Paul D. Lagasse, Master of Arts Thesis pp.42-45
7. Mark P. Sullivan, Dependable Engines: The Story of Pratt & Whitney (Reston, Virginia: American Institute of Aeronautics and Astronautics, Inc., 2008) pp.2-7
8. Antony L. Kay, Turbojet History and Development 1930-1960 p.98
9. Ibid. pp.99-100
10. Ibid. pp.101-102
11. Ibid. p.156
12. Bill Gunston, The Development of Jet and Turbine Aero Engines, (Sparkford, Nr Yeovil, Somerset, U.K.: Patrick Stephens, Ltd., 1995) p.153

Chapter Four

1. Edward H. Heinemann and Rosario Rausa, Ed Heinemann Combat Aircraft Designer (Annapolis, Maryland: Naval Institute Press, 1980) p.189
2. Rene Francillon and Edward Heinemann, Aerograph 5, Douglas A-3 Skywarrior (Arlington, Texas: Aerofax Inc., 1987) p.20

3. Ibid.
4. Bob Rahn with Rosario Rausa, Tempting Fate (North Branch, Minnesota: Specialty Press, 1997) p.105
5. Nick Williams "The Mayday Skyrays: Douglas Testing Division F4D-1s 1954-1959" AAHS Journal, Summer 1986, p.84
6. Tempting Fate pp.106-107
7. Ibid. p.107
8. Ibid.
9. Jim Stegman taped interview by Nick Williams, February 12, 1980
10. Tempting Fate pp.108-109
11. Ibid. pp.109-110
12. Arthur Pearcy, Flying The Frontiers: NACA and NASA Experimental Aircraft (Annapolis, Maryland: Naval Institute Press, 1993) pp.60-61

Chapter Five

1. Flight Handbook AN 01-40-FBA-1 pp.39-42
2. Ibid. pp.42-43
3. Ibid. p.44 and interview of LCDR Joel Premselaar USN (Ret.) by the Author on April 28, 2009
4. Flight Handbook p.47
5. Ibid. pp.47 & 50
6. Ibid. p.50 and interview of Maj. James R. Campbell, USMC (Ret.) by the Author on August 29, 2009
7. Naval Air Test Center Report, Flight Test Division, 11 December 1957 p.17
8. Ibid. p.19
9. Flight Handbook p.50
10. Ibid. pp.50-51
11. Ibid. pp.53-54
12. Ibid. pp.54- 58
13. Ibid. p.61
14. Ibid. pp.1 & 20, and Handbook Maintenance Instructions NAVWEPS 01-40-FBA-2 pp.3-1
15. Ibid. p.20A, and Handbook Maintenance Instructions NAVWEPS 01-40-FBA-2 pp.2-106J to 2-106AD
16. Handbook Maintenance Instructions NAVWEPS 01-40-FBA-2 pp.2-115 to 2-138
17. Ibid. pp.8-2 and 9-2 to 9-81
18. Ibid. pp.1-3 and 1-4, and Flight Manual F-102A & TF-102A T.O. IF-102A-1 pp.1-2, 1-3 and 5-9
19. Flight Manual F-102A & TF-102A pp.1-14, 1-15, 2-1 to 2-18A
20. Ibid. pp.2-19 to 2-22
21. Ibid. pp.6-1 to 6-7
22. Ibid. pp.2-29 to 2-32A and 6-6

Chapter Six

1. Nick Williams, "The Douglas Skyray in Fleet Colors, Part I" AAHS Journal, Summer 1978, p.129
2. Bill Gunston, Early Supersonic Fighters of the West, (London, Ian Allan Ltd., 1976) pp.230-231
3. Interview of Arland Nankivil by the Author. July 11, 2008
4. Nick Williams, "Silver Eagle Skyrays" The Hook, Spring 1978, pp.14-17
5. Letter to Author from James R. Campbell former VMF(AW)-314 pilot, dated March 23, 2007
6. Nick Williams, "The Douglas Skyray in Fleet Colors, Part I" AAHS Journal, Summer 1978, pp.136-137
7. Ibid. p.134.
8. Ibid. p.135
9. Nick Williams, "The Douglas Skyray in Fleet Colors, Part II" AAHS Journal, Summer 1981, pp.100-101
10. Ibid.
11. Letter from Gerard B. (Rod) Creagh, Jr. to the Author dated June 21, 2007
12. Nick Williams, "The Douglas Skyray in Fleet Colors, Part I" AAHS Journal, Summer 1978, p.137
13. Ibid. p.138
14. Ibid. pp.138-139
15. Nick Williams, "The Douglas Skyray in Fleet Colors, Part II" AAHS Journal, Summer 1981, pp.99-100
16. Ibid. pp.102-103
17. Ibid. pp.103-104
18. Ibid. pp.105-106
19. Ibid. p.108
20. Ibid. pp.109-110
21. Letter to the Author from Gary McClellan dated February 15, 2004
22. Nick Williams, "The Douglas Skyray in Fleet Colors, Part II" AAHS Journal, Summer 1981, p.107

Chapter Seven

1. Ibid. pp.110-111
2. Ibid. p.111
3. Ibid. p.112
4. Ibid. p.113
5. Nick Williams, "The Douglas Skyray in Fleet Colors, Part III" AAHS Journal, Summer 1985, pp.82-85

Chapter Eight

1. Tommy H. Thomason, U.S. Naval Air Superiority, Development of Shipborne Jet Fighters 1943-1962 (North Branch, Minnesota: Specialty Press, 2007) p.17
2. Nick Williams, "The Douglas Skyray in Fleet Colors, Part III" AAHS Journal, Summer 1985, pp.85-87
3. Elizabeth Babcock, History of the Navy at China Lake, Volume 3, Magnificent Mavericks (Washington, D.C.: Naval Historical Center, 2008) pp.413-416
4. Ibid. pp.425-442
5. Nick Williams, "The Douglas Skyray in Fleet Colors, Part III" AAHS Journal, Summer 1985, p.91
6. Ibid. p.90
7. Ibid. pp.88-89
8. Ibid. pp.90-91
9. Ibid. p.91
10. Ibid. pp.91-92
11. Nick Williams, "The Douglas Skyray in Fleet Colors, Part II" AAHS Journal, Summer 1981, p.114
12. Nick Williams, "The Douglas Skyray in Fleet Colors, Part III" AAHS Journal, Summer 1985, p.92
13. Letter to the Author from Phil Oestreicher dated March 17, 2007
14. Nick Williams, "The Douglas Skyray in Fleet Colors, Part III" AAHS Journal, Summer 1985, p.87-88

15. Wayne Mutza, Convair F-102 Delta Dagger (Atglen, Pennsylvania: Schiffer Publishing Ltd., 1999), pp.56-71
16. Ibid. pp.77-85
17. Ibid. pp.86-91
18. Ibid. pp.92-97
19. Ibid. pp.98-113 and 125-127

Chapter Nine

1. Bob Rahn with Rosario Rausa, Tempting Fate (North Branch, Minnesota: Specialty Press, 1997) p.149
2. Edward H. Heinemann and Rosario Rausa, Ed Heinemann Combat Aircraft Designer (Annapolis, Maryland: United States Naval Institute, 1980) pp.195-196
3. Steve Ginter, Naval Fighters Number Thirty-Five, Douglas F5D-1 Skylancer (Simi Valley, California: Steve Ginter, 1996) p.4
4. Rahn "Tempting Fate" p.145
5. Ibid. p.147
6. Ibid. pp.147-148
7. Ibid. p.148. Rahn may have recorded a full stall at 70 knots, but the Flight Handbook considered the stall speed 122 knots at takeoff weight with power off and 103 knots at landing weight with power off.
8. Steve Ginter, Naval Fighters Number Thirty-Five, Douglas F5D-1 Skylancer (Simi Valley, California: Steve Ginter, 1996) pp.10-11
9. Ibid. p.13
10. Ibid. p.14
11. Ibid. pp.13-14
12. Rahn "Tempting Fate" pp.149-150
13. Edward H. Heinemann, "A Veteran Designer's Thoughts on Fighters" Air International, October 1982, p.181
14. L. Stewart Rolls, David G. Koenig, and Fred J. Drinkwater III, "Flight Investigation of the Aerodynamic Properties of an Ogee Wing" NASA Technical Note, NASA TN D-3071, December 1965
15. Robert Godwin, Dyna-Soar—Hypersonic Strategic Weapons System, (Burlington, Ontario, Canada: Collector's Guide Publishing Inc., 2003) pp.224-226
16. Gene J. Matranga, William H. Dana, and Neil A. Armstrong, "Flight-Simulated Off-The-Pad Escape and Landing Maneuvers For a Vertically Launched Hypersonic Glider" NASA Technical Memorandum X-637
17. Steve Markman and Bill Holder, One-Of-A-Kind Research Aircraft, a History of In-Flight Simulators, Testbeds & Prototypes (Atgen, Pennsylvania: Schiffer Publishing Ltd., 1995) p.66
18. Utility Flight Handbook for Navy Model F5D-1 Aircraft, 1 July 1957
19. Bill Yenne, Convair Deltas: From SeaDart to Hustler (North Branch, Minnesota: Specialty Press, 2009) pp.119-125

Epilouge

1. Letter to Nick Williams from Gerald G. O'Rourke dated April 2, 1976
2. Jerry O'Rourke, "The Douglas Navy F4D Ford," AAHS Journal, Summer 1979, pp.125-131. This material also appeared in a slightly different form, "The F4D 'Ford': A Better Idea?" Naval Institute Proceedings, June 1986, pp.123—126
3. Letter to the Author from Phil Oestreicher dated August 27, 2008
4. Flight Report, "An Investigation of the Maneuvering Stability of the F-6A Airplane," P. F. Oestreicher, NATC Patuxent River, Maryland, dated 10 August 1966, and NATC Report of Test Results, "Investigation of the Transonic Flying Qualities of the F-6A Airplane," by P. F. Oestreicher, NATC Patuxent River, Maryland, dated 12 September 1966
5. Letter to Nick Williams from Phil Oestreicher dated January 5, 1982
6. Letter to the Author from Phil Oestreicher dated August 27, 2008
7. Letter to Nick Williams from George C. Watkins dated February 16, 1978
8. Edward H. Heinemann and Rosario Rausa, Ed Heinemann Combat Aircraft Designer (Annapolis, Maryland: United States Naval Institute, 1980) pp.184-185
9. Edward H. Heinemann, "A Veteran Designer's Thoughts on Fighters," Air International, October 1982, p.180
10. Nick Williams, "The X-Rays: Douglas' XF4D-1 Skyrays" AAHS Journal, Winter 1977, p.249
11. Bill Gunston, Early Supersonic Fighters of the West, (Shepperton, Surrey, United Kingdom: Ian Allan Ltd., 1976) p.215
12. Ibid. pp.215- 217
13. KLD Associates, "A Review of the Flight Characteristics of the F4D-1 and F7U-3 Aircraft with Particular Emphasis on Carrier Suitability" Final Report, July 7, 1986
14. Martin W. Bowman with Matthias Vogelsang, Lockheed F-104 Starfighter (Ramsbury, Marlborough Wiltshire, United Kingdom: The Crowood Press Ltd., 2000) p.21
15. Jerry O'Rourke, "The Douglas Navy F4D Ford" AAHS Journal, Summer 1979, pp.130-131

Index

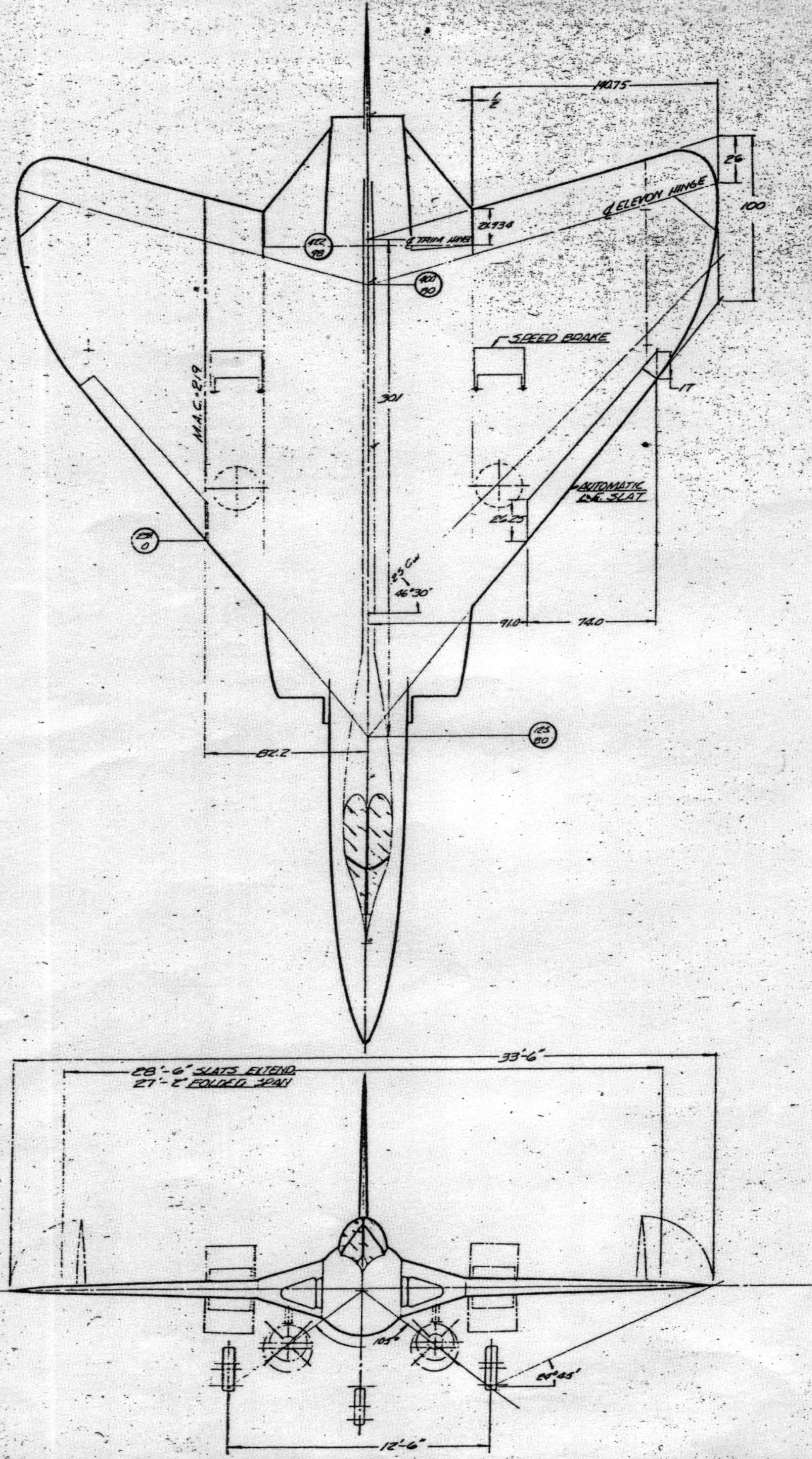

140.75
26
100
℄ ELEVON HINGE
21.734
℄ TRIM HINGE
400 130
SPEED BRAKE
M.A.C. 219
301
17
AUTOMATIC L.E. SLAT
26.25
25 C/4
46°30'
91.0
74.0
125 80
82.2
33'-6"
28'-6" SLATS EXTEND
27'-2" FOLDED SPAN
105°
12'-6"